INTERIM STORAGE OF SOLIDIFIED HIGH-LEVEL RADIOACTIVE WASTES

Panel on Engineered Storage
Committee on Radioactive Waste Management
National Research Council

NATIONAL ACADEMY OF SCIENCES
WASHINGTON, D.C. 1975

NOTICE: This report is part of a study undertaken under the aegis of the National Academy of Sciences–National Research Council (NAS–NRC) with the express approval of the Governing Board of the NRC. Such approval indicated that the Board considered that the problem is of national significance, that elucidation of the problem required scientific or technical competence, and that the resources of NAS–NRC were particularly suitable to the conduct of the project. The institutional responsibilities of the NRC were then discharged in the following manner:

The members of the study committee were selected for their individual scholarly competence and judgment with due consideration for the balance and breadth of disciplines. Responsibility for all aspects of this report rests with the study committee, to whom the Academy expresses its sincere appreciation.

Although the reports of study committees are not submitted for approval to the Academy membership or to the Council, each report is reviewed by a second group of appropriately qualified individuals according to procedures established and monitored by the Academy's Report Review Committee. Such reviews are intended to determine, *inter alia,* whether the major questions and relevant points of view have been addressed and whether the reported findings, conclusions, and recommendations arose from the available data and information. Distribution of the report is approved, by the President, only after satisfactory completion of this review process.

Sponsored by the U.S. Atomic Energy Commission, contract AT (11-1) - 3053

Library of Congress Cataloging in Publication Data

National Research Council. Committee on Radioactive Waste Management.

Interim storage of solidified high-level radioactive wastes.

"Sponsored by the U.S. Atomic Energy Commission, contract AT (11-1)–3053."
Includes bibliographical references.
1. Radioactive waste disposal. I. Title.
TD898.N345 1975 621.48′38 75.19146
ISBN 0-309-02400-5

Available from

Printing and Publishing Office
National Academy of Sciences
2101 Constitution Avenue, N.W.
Washington, D.C. 20418

Printed in the United States of America

COMMITTEE ON RADIOACTIVE WASTE MANAGEMENT

JOHN C. FRYE, Geological Society of America, *Chairman*
MANSON BENEDICT, Massachusetts Institute of Technology
W. KENNETH DAVIS, Bechtel Corporation
EARNEST F. GLOYNA, The University of Texas at Austin
KONRAD B. KRAUSKOPF, Stanford University
HERBERT M. PARKER, Consultant, Battelle Pacific Northwest Laboratories
ARTHUR M. PIPER, Private Consultant
RUSTUM ROY, The Pennsylvania State University
JOHN H. RUST, The University of Chicago
F. H. SPEDDING, Iowa State University
FRANK K. PITTMAN, U.S. Atomic Energy Commission, *Liaison*
CYRUS KLINGSBERG, NAS-NRC, *Executive Secretary*

PANEL ON ENGINEERED STORAGE

EARNEST F. GLOYNA, The University of Texas at Austin, *Chairman*
MANSON BENEDICT, Massachusetts Institute of Technology
W. KENNETH DAVIS, Bechtel Corporation
PETER T. FLAWN, The University of Texas at San Antonio
JOHN C. FRYE, Geological Society of America
*WARREN J. KAUFMAN, The University of California at Berkeley
J. FRANKLIN McCORMICK, University of Tennessee
ARTHUR M. PIPER, Private Consultant
MORTON SMUTZ, University of Florida
ROGER W. STAEHLE, The Ohio State University
RICHARD N. WHITE, Cornell University
WARREN F. WITZIG, The Pennsylvania State University
CYRUS KLINGSBERG, NAS-NRC, *Executive Secretary*

*Deceased

CONTENTS

PREFACE

This report responds to a request by the U.S. Atomic Energy Commission, dated November 7, 1972, to the Committee on Radioactive Waste Management (CRWM) of the National Academy of Sciences–National Research Council, for advice "concerning the AEC's concepts for the safe storage of high-level solidified waste for an interim period in an engineered storage facility at or near the surface of the earth."

A retrievable surface-storage facility (RSSF) for an interim period (not exceeding 100 years) is understood to be a provisional measure for use while procedures for long-term storage or ultimate disposal of wastes—as in deep geologic media—are being developed, proven, and adopted.

The prospective accumulation of radioactive wastes derived from nuclear-fission reactors presents a challenging, but resolvable, problem to modern society. Sufficient knowledge is available to proceed immediately with developing a plan for interim, retrievable, and safe surface storage. Immediate development of a plan for interim storage is a logical first step in a comprehensive system for the long-term management of the wastes.

The Panel on Engineered Storage of the Committee on Radioactive Waste Management has reviewed the three most promising of AEC's concepts—specifically, those based on solidified waste and involving storage in a water basin, an air-cooled vault, or a sealed storage cask. The Panel and the Committee on Radioactive Waste Management concluded that any of these concepts could lead to storing solidified high-level radioactive wastes generated by nuclear power reactors safely and economically. However, the sealed storage-cask concept, with its passive cooling system and multiple barriers against dispersal of waste, appears to offer substantial advantages. A sealed storage-cask facility can be designed and constructed to provide the necessary safety and reliability, with a minimum of operating attention.

In addition to critically reviewing the three retrievable surface-storage concepts mentioned, the report considers the relationship of such storage to ultimate disposal, to the segregation of transuranics, to considerations of safety and environmental integrity, to determination of optimum materials for construction, to container integrity, and to anticipated costs.

At the time the final draft of the report manuscript was under review, the Energy Reorganization Act of 1974 was enacted; this act transferred the AEC's responsibilities primarily to the Energy Research and Development Administration and to the Nuclear Regulatory Commission.

INTERIM STORAGE OF SOLIDIFIED HIGH-LEVEL RADIOACTIVE WASTES

CONCLUSIONS AND RECOMMENDATIONS

The Panel on Engineered Storage of the Committee on Radioactive Waste Management has analyzed the concepts proposed by the Atomic Energy Commission to hold solidified high-level radioactive wastes (produced by the nuclear power industry up to the year 2000) in retrievable surface storage for a period of up to 100 years. The Committee endorses the following conclusions and recommendations of the Panel:

1. The Panel has evaluated the available, pertinent information on management of high-level radioactive wastes that result from the operation of nuclear power reactors. It finds that retrievable surface storage can be developed that will (a) ensure public safety, through multiple barriers against dispersal of radionuclides and against exposure of persons to radiation; (b) minimize adverse effects on the environment; (c) permit operation within the necessary time scale; and (d) minimize the increase in nuclear power costs.

The Panel concludes that retrievable surface storage is an acceptable interim stage in a comprehensive system for managing high-level radioactive wastes.

2. The Panel has reviewed alternative methods of radioactive-waste

processing and storage. It concludes that, in principle, high-level radioactive wastes can be stored safely and economically in either sealed storage casks, in a water basin, or in an air-cooled vault.

However, *the Panel recommends an optimized version of the sealed storage-cask concept.*

3. The sealed storage-cask concept has these advantages: (a) The sealed cask is passive and thus requires a minimum of operating service or surveillance; (b) it has considerable design flexibility and thus can be readily modified according to operating experience; (c) it ensures containment of wastes under likely conditions of natural catastrophe or human intrusion, including sabotage or acts of war (however, the Committee did not study the possible effects of long-term political or social instability); (d) it is simple and economically acceptable; and (e) it requires a minimum amount of water, thereby increasing the number of potentially satisfactory sites.

The Panel recommends that a demonstration sealed storage-cask facility be built, to be ready for operation on production-run, solidified high-level radioactive wastes as soon as such wastes become available.

4. The Panel concludes that substantial protection is provided against the release of radioactive wastes by a series of five barriers. This series of barriers is designated as the *multiple barriers* concept and is a part of the sealed storage-cask plan. The barriers are (a) a solidified radioactive waste product that is chemically and mechanically stable; (b) canisters of a corrosion-resistant and mechanically stable material to enclose solidified waste and provide for its complete isolation from the environment during packaging, transportation, and early storage; (c) a sealed cask designed to retain mechanical and chemical integrity and to protect the canisters; (d) a concrete structure enclosing the sealed cask, which would provide mechanical protection and radiation shielding; and (e) the relative isolation of the site in a suitable geological region.

The Panel recommends that this concept of multiple sequential barriers be preserved and that the individual barriers be optimized to provide maximum reliability with no mutual deleterious interaction.

5. While the retrievable surface-storage concepts and associated technologies have been studied in considerable detail by the U.S. Atomic Energy Commission and its contractors, final specifications and designs have not been determined and considerable additional engineering research and development are necessary.

The Panel recommends the following subjects for prompt research and development, in a timely manner, with respect to both ordinary and accident conditions:

a. *optimization of the form of the solidified waste, striving for high stability, low solubility in water, and compact mechanical form;*

b. *study of performance of each of the cited barriers, including optimum materials for canister and overpack, and optimum specifications for concrete under the various environments of operation;*

c. *study of possible deleterious interactions between components of the waste and canister, overpack, or concrete—each extrapolated to about 100 years;*

d. *procedures for fabricating or processing reliable and economical components of the demonstration facility;*

e. *study of criteria for nondestructive testing of the components of the system;*

f. *procedures for filling, sealing, overpacking, and transporting the canisters;*

g. *specifications for the chemical and physical composition of the waste, the canister, and the concrete shield to maximize the long-term integrity of the facility;*

h. *procedures for monitoring thermal performance during all stages of the demonstration facility;* and

i. *procedures and techniques for the early detection of leaks or other failures of the canister, cask, concrete shield, or other structural members of the storage unit.*

6. The Panel concludes that the siting criteria suitable for each of the retrievable surface-storage concepts could be satisfied at many locations.

The Panel recommends that detailed studies of the site ecology, hydrology, and geology precede the establishment of a storage facility at the selected site or sites.

7. Any storage facility located on or near the surface of the earth is subject to damage from earthquake, tornado, sabotage, act of war, accidental air crash, etc.

The Panel recommends that precautions be taken in the siting and design of such facilities and in their operation so that the associated risks are minimized.

8. A storage facility on or near the surface of the earth might cause mutational changes because of the inadvertent intrusion of animal life, migrating birds, and insects.

The Panel recommends that steps be taken to make such a facility unattractive to plants and animals.

9. The Panel concludes that, although acceptable for an interim period, retrievable surface storage is not a substitute for ultimate disposal of high-level radioactive wastes.

Therefore, the Panel recommends an immediate and comprehensive assessment of sites in salt, shale, or other suitable geologic media for the disposal of solidified high-level radioactive wastes. This assessment should be followed by a demonstration repository at the earliest practicable time.

10. The Panel believes that separating the long-lived actinides from shorter-lived fission products may offer a great potential advantage in managing wastes over the long term.

The Panel recommends increased research on processes for economically separating and disposing of, or using, long-lived actinides.

INTRODUCTION

At the request of the Atomic Energy Commission (AEC), the National Academy of Sciences–National Research Council (NAS–NRC), through its Committee on Radioactive Waste Management (CRWM), accepted the charge to advise the Commission "concerning the AEC's concepts for the safe storage of high-level solidified waste for an interim period in an engineered storage facility at or near the surface of the earth." In response, CRWM constituted the ad hoc Panel on Engineered Storage to evaluate the technological aspects of storing solidified wastes in structures at or near the surface.

The expression, *retrievable surface-storage facility* (RSSF), or *engineered storage*, implies reliance on man-made systems maintained by appropriate surveillance, from which the stored waste can be recovered at a later date for transfer to long-term storage or other disposal. *Interim* indicates a period of up to 100 years.[1] Since the development of a system for long-term storage or ultimate disposal of high-level radioactive wastes must be pursued vigorously by AEC, the wastes may in fact be removed from interim surface storage well before 100 years.

SCOPE

This report reviews and evaluates the feasibility, safety, and estimated costs of several concepts for the interim surface storage of solidified high-level radioactive wastes that result from the civilian nuclear power program.

The main charge to the Panel was to recommend the best storage concept for further development from among those prepared by the AEC and its contractors. The Panel was also free to devise its own concept, if none of the AEC concepts appeared suitable from the overall point of view of feasibility, safety, or economy.

The alternatives considered by the Panel were presented only in conceptual forms, without the elaboration of detailed engineering design. An important function of the Panel in evaluating these alternative con-

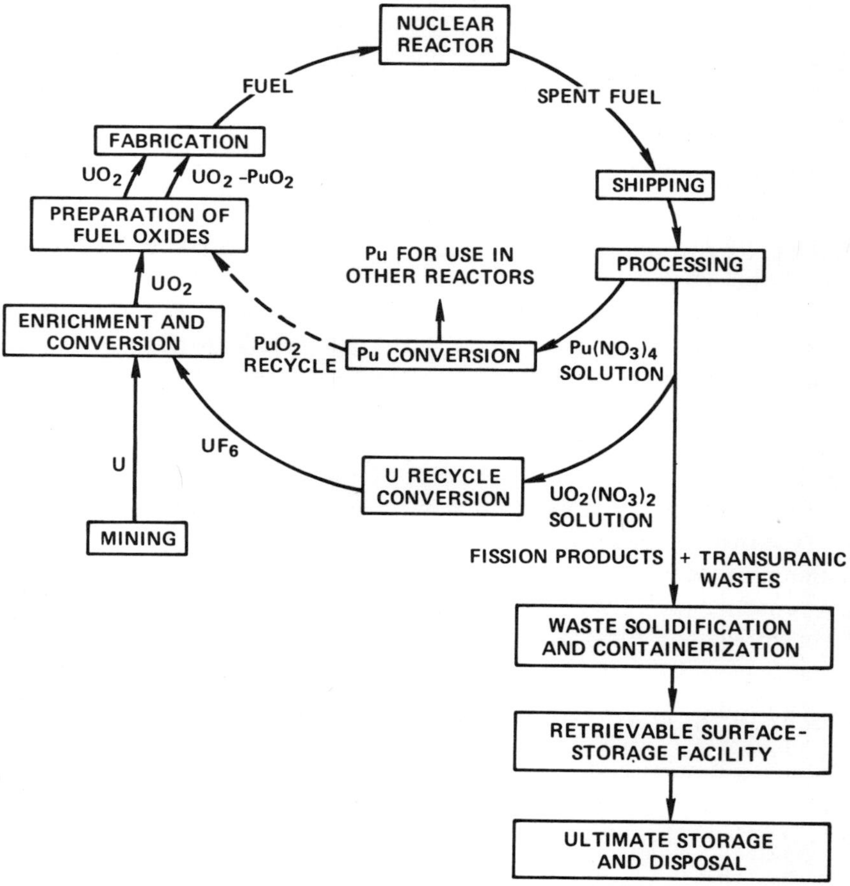

FIGURE 1
Typical nuclear fuel cycle for light-water reactor and waste.

cepts was to recommend where additional research was required to ensure reliability and adequate safety for the storage system.

The Panel gave primary attention to the physical problem of how best to store solidified waste in metal canisters at or near the surface of the earth. Attention was also given to the associated problems of safeguarding the wastes and minimizing danger due to acts of war. Another panel of the Committee on Radioactive Waste Management studied transportation problems. (See Appendix H.) The safety of each reprocessing facility is considered in its Safety Analysis Report.

In 1954, the Congress of the United States amended the Atomic Energy Act to permit the use of nuclear fuel in privately owned facilities to generate energy commercially. Figure 1 shows a typical nuclear-fuel cycle for light-water reactors. The U.S. Atomic Energy Commission established the policy of taking custody of all high-level nuclear wastes, of providing an acceptable management program, and ultimately of establishing storage or disposal programs capable of managing the wastes safely in perpetuity. Presently, the most promising practices for managing these wastes involve—

1. a requirement that the high-level wastes from nuclear fuel-reprocessing plants be solidified and then safely shipped to a federal repository within 10 years after reprocessing;
2. construction of an interim storage facility to store and isolate the waste from the biosphere until a permanent repository is established; and
3. determination of procedures for ultimate disposal and selection of sites for such disposal.

An economical means for safe management of wastes will require progressive technological advances in regard to concentrating and solidifying the wastes, packaging and transporting them, holding them for an interim period in a retrievable form, and finally, either long-term storage or ultimate disposal.

RETRIEVABLE SURFACE-STORAGE FACILITY (RSSF)

It is understood that the retrievable surface-storage facilities will be designed, constructed, and operated in accordance with AEC Regulations,[2] which require that wastes be stable, dry, and solid. Radioactivity from liquid or gaseous effluents released from the facility shall meet the applicable regulations.[3]

Deliveries of solidified wastes from chemical reprocessing plants can be expected to arrive at a U.S. government repository beginning about 1983, or earlier.

The design objectives of the RSSF must provide—

1. an early stage in a complete waste-management system;
2. a facility that can be operated until technological and social considerations permit the transfer of the wastes to long-term storage or to ultimate disposal;
3. components that are compatible with the next stage of the system, i.e., long-term storage or ultimate disposal;
4. safe operation, including the ability to remove heat, maintain canister integrity, and ensure contamination-free operations for the life of the facility;
5. assurance of the ability to retrieve the waste under foreseeable conditions; and
6. a facility that can withstand natural forces and make sabotage difficult.

The purpose of the RSSF is to confine waste in a limited space in high-integrity canisters. Cooling of the canisters may be passive or active. A wide variety of designs has been considered, e.g., the water basin concept (WBC), the air-cooled vault concept (ACVC), and the sealed storage-cask concept (SSCC).

RELATIONSHIP OF RETRIEVABLE SURFACE STORAGE TO LONG-TERM STORAGE OR ULTIMATE DISPOSAL

Management of radioactive wastes is complex and must be viewed as a complete system. (Figure 2 illustrates this complexity.) In the context of this report, *ultimate disposal* (as distinct from *long-term storage*) means isolating the radioactive wastes permanently in a form and manner that precludes retrieving them. Of the means proposed to date, such disposal would be best accomplished by placing the wastes away from the biosphere in an appropriate geologic and hydrologic environment.

As stated in a preceding section, the RSSF concept is an adequate interim means of storage. It permits orderly technological and social progress toward a truly comprehensive plan for ultimate management of high-level radioactive wastes. Long-term storage, or ultimate disposal, in a dry salt or shale formation has already been shown to be feasible.[4–9]

Salt deposits are widely distributed in the United States, some in seismically stable regions.[10] Salt has desirable plastic and thermal properties. The existence of extensive, thick salt deposits in a given

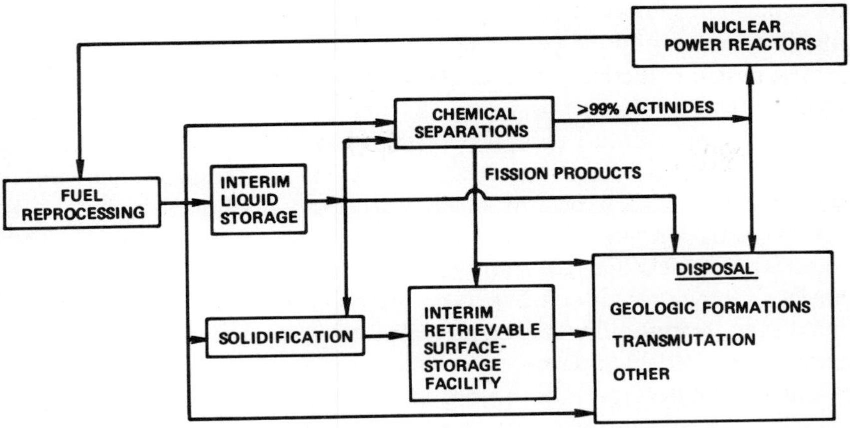

FIGURE 2
Options for management of high-level wastes.

area signifies the virtual absence of circulating water in that part of the geologic sequence during the time since the salt deposit was formed. Salt is easy to mine, and its geologic occurrence permits a wide range of depths for burial.

Battelle Pacific Northwest Laboratories has studied alternative concepts for ultimate disposal, including placement in polar ice caps, in interplanetary space, in deep rock formations other than salt, and in selected deep ocean trenches.[11] Each of these concepts has advantages and disadvantages.

Management of high-level radioactive wastes in other countries continues to move ahead. For example, in the Federal Republic of Germany, solidified wastes from fuel reprocessing plants soon will be placed in rock-salt formations.[12] Great Britain plans to transform its liquid wastes to borosilicate glass, pour this glass into steel canisters, and store the canisters in a cooled water basin.[13] France will use surface storage, but has chosen an air-cooled system for pilot studies.[14]

VOLUMES OF HIGH-LEVEL WASTES

Estimates of volumes of radioactive wastes necessarily reflect the state of the economy and the willingness of the public to accept nuclear reactors. One estimate suggests that the installed nuclear electric capacity of the United States at the end of 1980, 1990, and 2000 may be 134, 504, and 1,200 gigawatts respectively.[15] Based on present methods of waste processing, reactor capacity of these magnitudes might produce an equivalent of about 18,000 cubic yards (13,800 m^3) of solidified high-level wastes by the year 2000. These wastes (after 10 yr of decay) would contain an anticipated 53,000 MCi of activity and produce heat at the rate of 195 megawatts.

Two separate factors must be considered in protecting man from the hazards of radioactive materials. In the first place, shielding must be provided to control exposure to penetrating radiation, principally gamma radiation, but with some neutron component. In the second place, the material must be confined so that it does not escape to the environment, where it could be taken up by man through ingestion or inhalation. In this mode, the radionuclides that emit alpha or beta particles are very significant.

During the first several centuries of radioactive waste management, the type and degree of protection needed will be dictated by the relatively large amounts of penetrating radiation from fission products; but, for thousands of years after the fission products have become innocuous by decay, protection against inhalation or ingestion of the comparatively small amounts of alpha-emitting transuranium elements will become paramount. Consequently, protection against radiation from radioactive wastes requires both containing the wastes to prevent their dispersal and storing the containers so that penetrating neutron and gamma radiation does not reach man or other living matter in unacceptable doses.

Historically, both surface and subsurface facilities have been used effectively to shield, immobilize, and isolate large quantities of highly radioactive materials. A significant factor in regard to storage and ultimate disposal is that, within 800–1,000 years, almost all the fission-product species will have decayed to an activity level that requires no confinement. Thus, managing high-level wastes that contain essentially no transuranics is within man's historical ability. On the other hand, if the concentration of transuranics is appreciable, decay to an innocuous level could require hundreds of thousands of years, and man has no experience with containing or isolating any substance over a term that long.

FUEL REPROCESSING AND WASTE SOLIDIFICATION

THE FUEL CYCLE IN GENERAL

For the next 30 years, most nuclear power reactors in the United States will be of the present light-water type. Such reactors use a ceramic fuel mixture of uranium-235 and uranium-238 in which the uranium-235 isotope has been concentrated to the level of 2–3 percent, rather than the naturally occurring 0.7 percent. The operating reactor generates heat by fission of the nuclear fuel, with most of the ^{235}U being consumed and some (~1%) of the ^{238}U being converted to plutonium. A major fraction (over 70%) of this plutonium is in the form of plutonium-239 and plutonium-241, which are fissionable isotopes and can be used as fuel in place of ^{235}U, in both present reactors and future fast breeder reactors. Very small amounts of other transuranic elements are produced in addition to the plutonium.

When only about 3 percent of the uranium has undergone fission, the fission products that accumulate within the fuel seriously interfere with the operation of a reactor. Therefore, about once a year, a set of fuel elements in the reactor is withdrawn and replaced by a fresh set. From the spent fuel, the fission products and associated transuranics are separated by chemical reprocessing; once separated, they constitute the wastes whose preparation for storage and placement in storage are the particular concerns of this report.

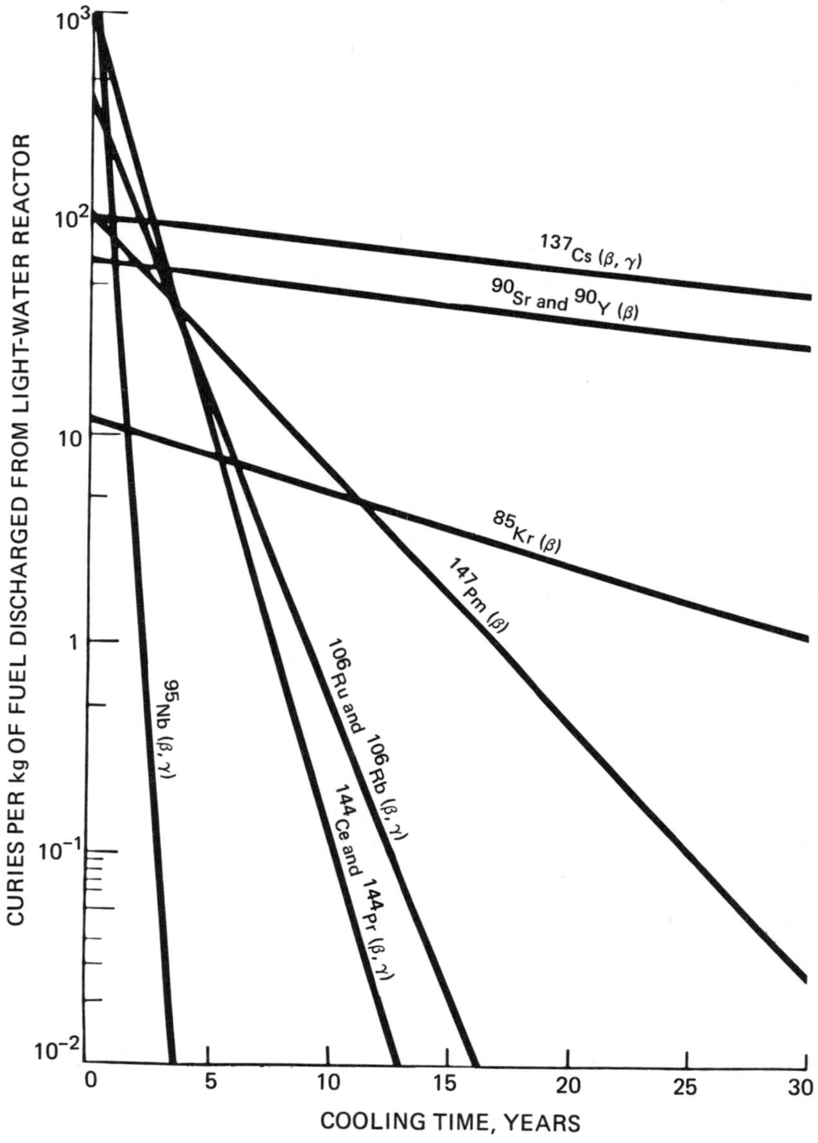

FIGURE 3

Isotopic activity (Ci/kg of U) of fission products in waste resulting from a kilogram of uranium charged to a light-water reactor with a burn-up of 33,000 MWd/metric ton. [Data from: *Siting of Fuel Reprocessing Plants and Waste Management Facilities,* ORNL-4451, Oak Ridge National Laboratory, Oak Ridge, Tenn. (July 1970) (Compiled and edited by the staff) and Bell, M. J., *ORIGEN–The ORNL Isotope Generation and Depletion Code,* ORNL-4628, Oak Ridge National Laboratory, Oak Ridge, Tenn. (May 1973).]

As the nuclear fuel undergoes fission, the nucleus of the atom divides into two or more parts with the release of heat and one or more neutrons. About 97 percent of the products of such fission fall into two groups: The light group has mass numbers from 85 to 104, and the heavy group has mass numbers from 130 to 149. The newly formed fission products are elements from nearly all groups in the periodic table. Most of them are highly radioactive, emitting beta particles and gamma rays. Their half-lives range from a fraction of a second to many years. Figure 3 shows the radioactivity levels of certain fission products as a function of time.

Because the spent fuel elements that are freshly withdrawn from a reactor are intensely radioactive, current practice is to store them under water at the reactor site to permit them to decay prior to shipping them to a reprocessing plant in a specially designed shipping cask. After 10 years of decay, the remaining hazard from fission products is dominated by two radioisotopes, each with a half-life of about 30 years, i.e., strontium-90 and cesium-137 (Figure 3). After about 350 years, however, the hazard from these two radioisotopes will have decreased by more than three orders of magnitude. From that time on, the hazard comes increasingly from the very small amount of long-lived plutonium and other alpha-emitting particles that remain (Figure 4).

FUEL REPROCESSING

At the present time, there are no commercial fuel-reprocessing plants in operation in the United States, but two* are under construction: Nuclear Fuel Services at West Valley, New York (temporarily out of service while undergoing modification); and Barnwell Nuclear Fuel Plant at Barnwell, South Carolina. Both of these plants will use modifications of the Purex solvent extraction process to separate the fission products and to recover the uranium and plutonium. This process has been used successfully in AEC production facilities for more than 20 years. It is of some concern to the Panel and the Committee that there will be no reprocessing plants in operation in the United States prior to mid-1976.

When fuel elements arrive at a reprocessing plant, they are carefully removed from their shipping casks, inspected, chopped into convenient-sized pieces, and dissolved in nitric acid. Although processing details differ at the two plants, each will produce a high-level waste product that must be stored for a very long time. Plutonium is highly toxic when inhaled and has a much longer half-life than fission products

*The General Electric Co. has withdrawn its Application for a License to Operate the Midwest Fuel Recovery Plant at Morris, Illinois.

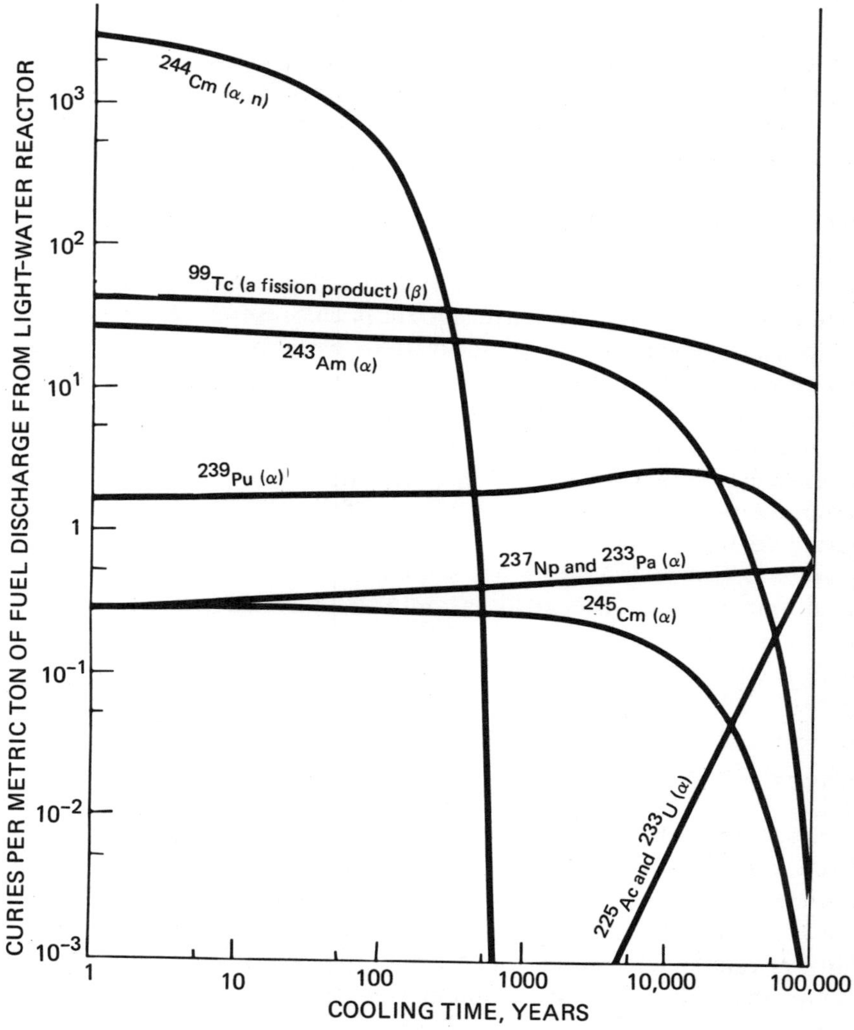

FIGURE 4

Isotopic activity (Ci/t of U) of actinides and their daughters in waste resulting from a metric ton of uranium charged to a light-water reactor, with a burn-up of 33,000 MWd/metric ton. [Data from: *Siting of Fuel Reprocessing Plants and Waste Management Facilities,* ORNL-4451, Oak Ridge National Laboratory, Oak Ridge, Tenn. (July 1970) (Compiled and edited by the staff) and Bell, M. J., *ORIGEN–The ORNL Isotope Generation and Depletion Code,* ORNL-4628, Oak Ridge National Laboratory, Oak Ridge, Tenn. (May 1973).]

(about 25,000 yr for ^{239}Pu as compared to about a 30-year maximum for the primary fission products). As a significant fraction (about 0.5%) of the plutonium is not recovered from the wastes during reprocessing, the level of radioactivity of that waste over the long term (beyond 1,000 yr) is the result of the plutonium and other transuranic elements, and not of the fission products. If plutonium and other transuranic elements could be almost completely removed from the waste, the storage time could be reduced to about 1,000 years. (See Appendix A—Separation of Transuranics.)

High-level wastes have rates of heat generation (decay rates) that determine the practicable modes of their handling, processing, and storage. Figure 5 shows the total heat generated, as well as that contributed by the more important families of isotopes as a function of time. The units are in watts per kilogram of fission product and actinide elements in high-level wastes from a typical light-water reactor in which fuel has been exposed to an average burn-up of 33,000 MWd per metric ton at an average specific power of 30 MW per metric ton. Ten years after fuel reprocessing, the heat-generation rate of the wastes is approximately 25 W/kg, which is about 1/20 of the rate at the time of reprocessing. To effect another tenfold reduction in decay heat, one must wait for about another 100 years. With this trend in heat decay, and in view of the difficulties in handling high-level wastes, it is convenient to defer their handling and further processing for at least several years.

WASTE SOLIDIFICATION

Under current AEC regulations, permanent custody by the federal government is required for all high-level radioactive wastes generated by commercial nuclear fuel-reprocessing plants. These wastes must, within 5 years after reprocessing, be converted to solids that are chemically, thermally, and radiologically stable; they must be placed in sealed containers and, within 10 years, shipped to a federal repository (*Federal Register*, 35(222):17530, November 14, 1970). A reprocessing plant may produce any type of solid that meets the above criteria.

The solidified wastes will not be of constant composition and may vary appreciably from time to time, even from the same reprocessing plant. Consequently, their potential corrosivity is difficult to assess. In the 1960's, "calcine" evolved as an acceptable form of solidified waste, on the expectation that it would be disposed of in salt formations. Much more desirable forms of the solidified waste are now known or conceivable, depending on the concept selected for storage

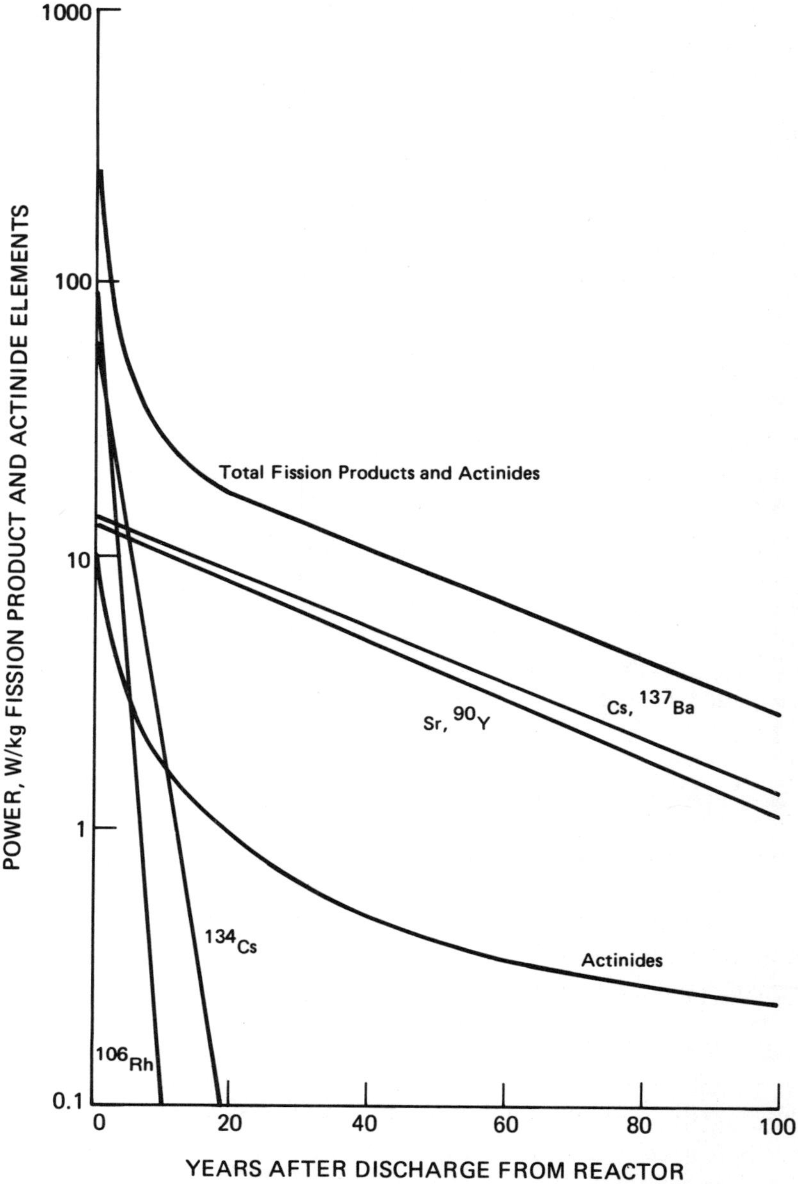

FIGURE 5
Rate of heat generation by wastes from pressurized water reactor (data from ORNL-4431).

and disposal. Phosphate and silicate glasses incorporating the radionuclides have been designed and studied; these are several orders of magnitude less soluble than "calcine." The design of the total RSSF system must remain flexible until the chemical composition and physical form of the solidified waste can be determined.

Table 1 gives characteristics of wastes in the form of calcine or borosilicate glass. Table 2 gives the composition of high-level liquid-waste solutions prior to conversion to solid form.

TABLE 1
Characteristics of Typical Solidified High-Level Waste

	Calcine	Borosilicate Glass
Form	Granular	Monolithic
Description	Amorphous	Black glass
Fission product oxide, mol %	Up to 90	Up to 30
Bulk density, g/cm^3	1.0–1.7	3.0–3.4
Thermal conductivity, $W/(m^2)$ ($^\circ C/m$)	0.15–0.4	0.8–1.2
Leachability at 25 °C, g/cm^2-day	$1\text{–}10^{-1}$	$10^{-5}\text{–}10^{-7}$
Relative surface/volume ratio[a]	10^4	1
Hardness	Soft	Very hard
Frangibility	Crumbly	Brittle
Residual nitrate, wt %	≤4.0	<0.005
Volume, liter/1,000 MWdth	1.5–5.0	1.5–5.0
Maximum stable temperature, °C	600	600[b]

SOURCE: Summary of data presented to NAS-NRC Panel on Engineered Storage by J. L. McElroy, Battelle Memorial Institute, Battelle Pacific Northwest Laboratories, Richland, Washington.
[a]Based on monolith glass samples approximately 25 × 250 cm long and calcine particles averaging between 10 and 100 μ (0.001–0.01 cm) in diameter.
[b]Onset of devitrification.

DELIVERY OF CANISTERS

Under current federal regulations, deliveries of radioactive wastes to U.S. government repositories may begin about 1983. (See Table 4.) In view of the limited time remaining for construction, one or more interim surface-storage facilities should be authorized at the earliest date possible. The relative merits of one site *vs.* multiple sites are discussed in *Management of Commercial High-Level and Transuranium-Contaminated Radioactive Waste* (AEC Environmental Statement, WASH-1539, September 1974, draft).

TABLE 2

Typical Materials in High-Level Liquid Waste

	Grams/t from Reactor Type[a]		
Material[b]	LWR[c]	HTGR[d]	LMFBR[e]
Reprocessing chemicals			
Hydrogen	400	3,800	1,300
Iron	1,100	1,500	26,200
Nickel	100	400	3,300
Chromium	200	300	6,900
Silicon	–	200	–
Lithium	–	200	–
Boron	–	1,000	–
Molybdenum	–	40	–
Aluminum	–	6,400	–
Copper	–	40	–
Borate	–	–	98,000
Nitrate	65,800	435,000	244,000
Phosphate	900	–	–
Sulfate	–	1,100	–
Fluoride	–	1,900	–
Subtotal	68,500	452,000	380,000
Fuel product losses[f,g]			
Uranium	4,800	250	4,300
Thorium	–	4,200	–
Plutonium	40	1,000	500
Subtotal	4,840	5,450	4,800
Other transuranium elements[g]			
Neptunium	480	1,400	260
Americium	140	30	1,250
Curium	40	10	50
Subtotal	660	1,440	1,560
Other actinides[g]	<0.001	20	<0.001
Total fission products[h]	28,800	79,400	33,000
TOTAL	103,000	538,000	419,000

SOURCE: U.S. Atomic Energy Commission, *High-Level Radioactive Waste Management Alternatives,* WASH-1297 (May 1974).

[a]Water content is not shown; all quantities are rounded.

[b]Most constituents are present in soluble, ionic form.

[c]^{235}U-enriched PWR, using 378 liters of aqueous waste per metric ton, 33,000 MWd/t exposure. [Integrated reactor power is expressed in megawatt-days (MWd) per unit of fuel in metric tons (t).]

[d]Combined waste from separate reprocessing of "fresh" fuel and fertile particles, using 3,785 liters of aqueous waste per metric ton, 94,200 MWd/t exposure.

[e]Mixed core and blanket, with boron as soluble poison, 10% of cladding dissolved, 1,249 liters per metric ton, 37,100 MWd/t average exposure.

[f]0.5% Product loss to waste.

[g]At time of reprocessing.

[h]Volatile fission products (tritium, noble gases, iodine and bromine) excluded. Not all these are in fact considered volatile, e.g., ^{129}I is assumed to be present and is an important long-range nuclide.

RETRIEVABLE SURFACE-STORAGE CONCEPTS

Engineering studies under way for a retrievable surface-storage facility (RSSF) are intended to satisfy requirements for high-level radioactive waste storage until a method for ultimate disposal is established. These studies are based on shipment of solidified high-level radioactive wastes to a central federal repository after storage of up to 10 years at a reprocessing plant. Standardized shipping canisters are assumed; therefore, the receiving facility for all RSSF storage systems would be the same. Only the facilities necessary to prepare the canisters for storage would vary.

The preliminary RSSF concepts given serious consideration in this report are—

1. sealed storage-cask concept—shielded (SSCC—Shielded),
2. sealed storage-cask concept—thick wall (SSCC—Thick Wall),
3. sealed storage-cask concept—unshielded (SSCC—Unshielded),
4. air-cooled vault concept (ACVC),
5. water basin concept (WBC), and
6. water basin concept—overpack (WBC-OP).

Table 3 compares salient technical and economic characteristics of these six concepts. The descriptions of the several concepts are based

TABLE 3
Comparison of Retrievable Surface-Storage Facility Concepts

	Sealed Storage-Cask Concepts			Air-Cooled Vault Concept (ACVC)	Water Basin Concepts	
	SSCC–Shielded	SSCC–Thick Wall	SSCC–Unshielded		WBC	WBC-OP
Cooling method	Air natural circulation	Air natural circulation	Air natural circulation	Air natural circulation	Water-forced circulation and cooling towers	Water-forced circulation and cooling towers
Confinement method	2″ Wall sealed carbon-steel cask welded closure	14″ Wall sealed carbon-steel cask welded closure	8″ Wall sealed carbon-steel cask welded closure	0.5″ Wall carbon steel overpack welded closure	Water-filled steel-lined concrete basin	0.5″ Wall, stainless steel overpack welded closure water-filled basin
Biological shielding	38″ Thick concrete sleeve	11″ Thick concrete sleeve	None (distance)	Concrete vaults	Water basin	Water basin
Storage concept	Vertical outdoor on 20-foot centers	Vertical outdoor on 20-foot centers	Horizontal outdoor 20 ft apart	Vertical indoor	Vertical indoor	Vertical indoor
Maximum normal operating temperature (°F)						
Canister (outer surface)	470[a]	450[a]	455[a]	620[b]	<128	484
Overpack (outer surface)	245[a]	170[a]	180[a]	400[b]	NA	127

Concrete (inner surface)	155[a]	122[a]	NA	200[b]	NA	NA
Dose rates (mrem/h)						
Outside surface	2 (concrete sleeve)	2 (concrete sleeve)	14,000	<2 (top of deck)	<1 (water surface)	<1 (water surface)
Within array	<4	<4	10,000	NA	NA	NA
Resources (to year 2010)						
Land (acres)	1,100	1,100	2,300	100	100	150
Water (gal/yr)	75×10^6	75×10^6	75×10^6	75×10^6	$1,500 \times 10^6$	$1,500 \times 10^6$
Carbon steel[c] (tons)	0.28×10^6	2.3×10^6	0.98×10^6	0.15×10^6	0.10×10^6	0.22×10^6
Stainless steel (tons)	42	42	42	42	5,600	45,000
Concrete (yd^3)	2.7×10^6	1×10^6	0.09×10^6	0.34×10^6	0.1×10^6	0.14×10^6
Personnel	330	330	330	280	250	330
Power (kWe)	2,500	3,000	3,000	2,100	10,400	10,000
Capital[d] (in 1973 dollars)	900×10^6	$2,600 \times 10^6$	$1,130 \times 10^6$	460×10^6	280×10^6	740×10^6
Cumulative operating costs (in 1973 dollars through year 2010)	220×10^6	200×10^6	190×10^6	165×10^6	170×10^6	220×10^6

SOURCE: *Retrievable Surface Storage Facility: Alternative Concept Engineering Studies,* ARH-2888, Rev., Atlantic Richfield Hanford Company, Kaiser Engineers (July 1974).

[a] 80 °F ambient air temperature.
[b] 110 °F ambient air temperature.
[c] Including structural and reinforcing steel.
[d] Including cost of steel overpacks and concrete shields.

on the preliminary data from AEC and are not to be construed as final conceptual designs. The shielded sealed storage-cask concept (SSCC–Shielded), the air-cooled vault concept (ACVC), and the water basin concept with overpack (WBC-OP) were deemed to be the best alternatives and are compared in detail in the following sections. The remaining three concepts (i.e., concepts 2, 3, and 5 above, described in Appendices B, C, and D) are variants of the SSCC–Shielded and WBC-OP and are not considered further in this report.

According to current design, solidified waste will be packaged in a standard stainless steel canister of about 12-inch (30-cm) inside diameter that is 10 feet (3 m) long and contains 6.3 cubic feet (0.18 m^3) of waste from reprocessing 3.15 metric tons of fuel and generating approximately 3.25 kilowatts of decay heat at 10 years of age. (Actually, a decay heat of 5 kW per canister was used as a design basis.) The radioactive waste in each canister results from the generation of approximately 800 million kilowatt hours of electricity.

Table 4 is a condensed schedule of waste delivery and of attendant cumulative decay heat. The estimated 75,000 standard canisters will accommodate approximately all the high-level wastes expected to be produced in the United States by the year 2000. The last waste shipment to the RSSF will take place within 10 years of its generation, or by 2010. The RSSF is to be designed for up to 100 years of storage with continued surveillance and monitoring. After providing for the initial receiving and handling facilities by around 1981, a modular construction program is anticipated to allow postponement of capital expenditures until added storage facilities are required.

The SSCC–Shielded is the forerunner of the SSCC considered in detail in this section. While both designs provide sufficient shielding to permit access to the storage area, the thick-wall design [14-in. (36-cm) thick wall and an 11-in. (28-cm) thick concrete sleeve] was deemed to be

TABLE 4
Projected Waste Canister Delivery to the RSSF[a]

Year	No. Waste Canisters Shipped: Per Year	Cumulative	Heat Inventory Cumulative (MW)
1983	15	15	0.03
1990	800	3,100	8
2000	3,100	21,900	74
2010	7,300	75,000	195
2100	0	75,000	30

SOURCE: Table 4.1, ARH-2888 Rev., page 4.2.
[a]100 years of storage has been assumed.

considerably less than optimum with respect to shielding efficiency and use of materials. The SSCC–Shielded, using a 2-inch (5-cm) cask wall and a 38-inch (97-cm) concrete sleeve, is a more optimal design.

The unshielded sealed storage-cask concept [8 in. (20-cm) of steel] would provide limited shielding, so that all operations within the storage area would have to be conducted from shielded vehicles. This concept did not seem to offer any advantage relative to the SSCC and was, therefore, dropped from further consideration.

The WBC is identical to the WBC-OP except for the omission of the thin-wall steel overpack. It was generally agreed that each concept should offer at least two effective confinement barriers; thus, since the water itself could not be relied upon as one of these barriers, the WBC without an overpack is not an acceptable concept from the viewpoint of safety.

In summary, the three concepts selected for further examination are the SSCC–Shielded, ACVC, and WBC-OP. For purposes of comparison, a summary description of the current method for storing high-level wastes as a liquid in steel tanks is included in Appendix E.

SEALED STORAGE-CASK CONCEPT–SHIELDED

In the shielded sealed storage-cask concept, each waste canister, after being received at the waste facility, is placed within an overpack, i.e., a low-carbon steel cask, 2 inches (5-cm) thick, with a 15-inch (38-cm) inside diameter, and sealed by welding. The steel storage cask is surrounded by an annular concrete gamma-and-neutron shield 38 inches (97-cm) thick, with a 31-inch (79-cm) inside diameter and transported to an outdoor storage area where it is mounted vertically on a concrete base, as schematically depicted in Figure 6. Casks are set in a square array, approximately 20 feet (6 m) between centers (Figure 7). Heat is dissipated by natural convective air flow through the 6-inch (15-cm) wide annulus between the steel cask and the concrete shield.

Salient Features

With heat being generated at the design rate of 5 kW* from 10-year old waste, air will flow through the annulus at 3 ft/s (0.9 m/s) and will experience a temperature rise of 35 °F (20 °C). With inlet air at 80 °F

*In Appendix F (and in some other sections of this report), an SSCC with a 10- and an SSCC with a 20-kilowatt heat load are also considered as other variables in the overall optimization of this concept.

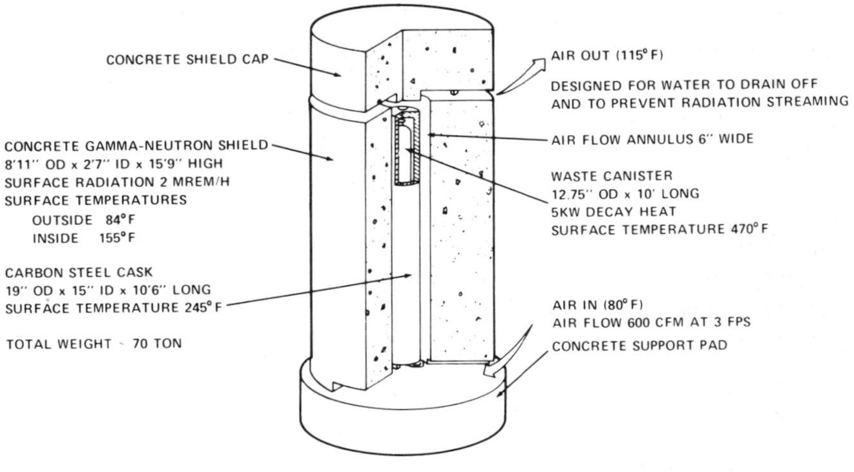

FIGURE 6
Diagram of a preliminary AEC storage unit for the sealed storage-cask concept.

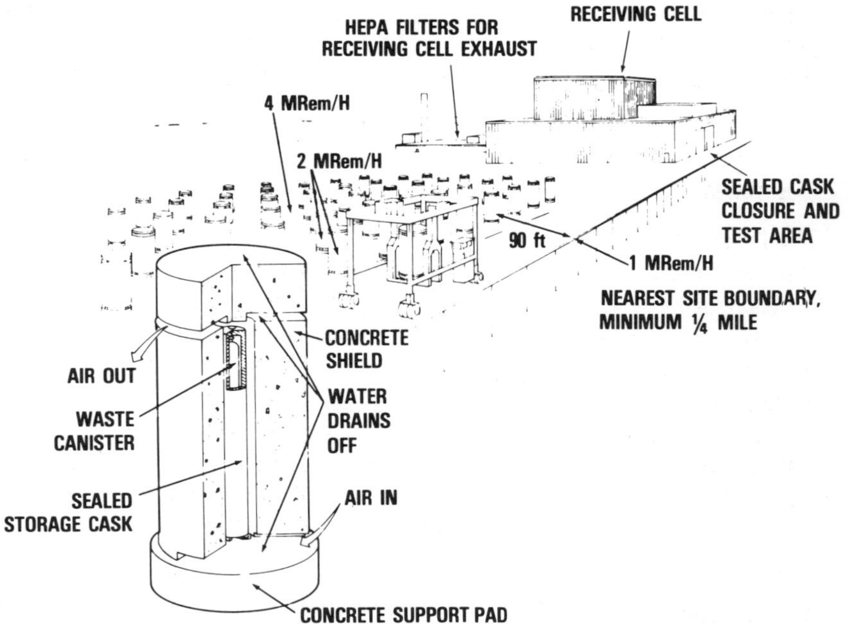

FIGURE 7
Diagram of a preliminary AEC sealed storage-cask concept facility.

(27 °C), maximum temperatures under normal operating conditions are estimated to be 155 °F (68 °C) for the concrete, 245 °F (120 °C) for the carbon steel cask, and 470 °F (245 °C) for the stainless steel canister; these temperatures are well below the limiting design temperatures of 500 °F (260 °C) for the concrete, 550 °F (290 °C) for carbon steel, and 800 °F (430 °C) for stainless steel.

The radiation dose rates are estimated to be less than 4 mrem/h anywhere in the storage array, 2 mrem/h at the outside of the concrete shield, and less than 1 mrem/h 90 feet (27 m) from the array of casks. In the 6-inch (15-cm) annulus between the steel cask and concrete shield, the gamma dose rate will be about 20,000 R/h, which is enough to provide a lethal dose in a few minutes to small mammals that might stray into the space. Because of the very high dose rate at its surface, the steel cask will have to be welded shut with remotely controlled equipment and be transported from the welding station in its concrete shield or in a heavily shielded vehicle. Neutron radiation from the cask will slightly activate the dust conveyed past it by the cooling air, but this effect is calculated to be minimal.

Corrosion of the steel cask, expected to be at an average rate of less than 0.010 inch (0.25 mm) per year,[16] is probably slow enough to ensure integrity of the steel cask for over 100 years, but this rate needs experimental verification at temperatures, chemical environments, and radiation levels representative of storage conditions. Special attention should be given to minimizing corrosion. (See "Reliability of Materials," pp. 41ff.)

Resources

The largest part of the cost of the SSCC–Shielded facility is for the steel casks and concrete shields. In order to store the 75,000 canisters expected to be needed by the year 2010, 180,000 tons of steel will be used for canisters and casks, 100,000 tons of steel will be needed for other purposes, and 2.7 million cubic yards (2.1 million m^3) of concrete will be required for shields and pads.

Personnel requirements will be approximately 90 in 1983, 330 in 2010, and 80 thereafter. In 2010, water usage will be about 75 million gallons (280,000 m^3) per year, and electrical usage 2,500 kilowatts. Eleven hundred acres will be required to store 75,000 casks.

Technology

Heat removal from the wastes by conduction through the cask and by radiation and convection from the outer surface of the cask depends on elementary principles of fluid flow and heat transfer that are so well known as to require no further research or development. Some attention should be given to possible mechanisms that might obstruct air flow (such as small animals, birds, concrete spalls, or windblown debris), to means for detecting and clearing obstructions, and to drainage of water.

Innovative engineering will be required to extend existing welding and nondestructive testing techniques to remote closure of 2-inch-thick steel casks. Tests are needed to establish that the 2-inch-thick steel cask will remain leak-tight under anticipated conditions of temperature and radiation during the required 100-year storage period.

The thicknesses chosen for the steel cask [2 in. (5 cm)], cooling annulus [6 in. (15 cm)], and concrete shield [38 in. (97 cm)] are the result of a preliminary optimization study, which should be extended before final dimensions for the storage units are set.

Summary Evaluation

The flexibility of the sealed cask concept is one of its most attractive features. Facilities can be readily modified as the nature of the wastes, the dimensions of the waste canisters, or the total number of canisters change.

Other attractive features of this concept are its passive character and the need for a minimum of operating surveillance once the wastes are enclosed in the steel casks and placed in the concrete shields. This storage concept can tolerate long interruptions in its power supply because cooling is not dependent on either circulating water or on forced-draft ventilation. Leakage, should any occur, can be readily detected and located by radiological scanning of the storage area.

This storage system is practically immune from damage caused by earthquake, tornado, impacting aircraft, or sabotage because of the massive concrete shields. Even if a cask and shield were buried to a depth of 5 feet (1.5 m), temperature would reach only 600 °F (315 °C) after 10 days, 1400 °F (760 °C) after 5 weeks, and an asymptotic temperature of 1700 °F (930 °C) (Appendix F), which is well below the melting point of steel.

This concept places total reliance on the integrity of the steel cask over a 100-year period. Like other concepts, it requires continual monitoring. If leaks should occur, the defective cask could be identified, overpacked, and returned to storage. Also, the concrete shields could be replaced if conditions so warranted.

WATER BASIN CONCEPT WITH OVERPACK

In the water basin concept with overpack, the waste canisters are placed in 0.5-inch (1.3-cm) thick, 16-inch (40-cm) outside diameter, stainless steel overpacks that are sealed by welding (Figure 8). These canisters and their overpacks are stored in stainless steel-lined concrete basins. Each basin has a nominal capacity of 500 canisters placed in double-tiered racks. The basins are filled with demineralized water, which cools the canisters and provides transparent radiation shielding, as well as further confinement should a leak develop in both canister and overpack.

Salient Features

Heat, removed by continuous pumped circulation of the basin water in direct contact with the overpacks, is transferred to a secondary cooling loop and rejected to the atmosphere from standard, open, forced-draft cooling towers located just outside the storage building. Figure 9 is a diagram of the system.

Each basin is equipped with a pump and heat-exchanger system to maintain basin water temperature below 120 °F (50 °C) and the canister surface temperature below 484 °F (250 °C). The overpack surface temperature is less than 127 °F (60 °C). Water purity is maintained at less than 10 milligrams of chloride per liter by passing a portion of the circulated cooling water through a filter and a demineralizer. The radiation dose rate at the surface of the water in the basin is estimated to be less than 1 mrem/h.

The WBC-OP uses a cooling system that requires continuous surveillance and occasional equipment maintenance. Equipment for close control over water chemistry is needed to minimize corrosion of overpacks and basin liners. The basins and liners provide physical protection for storage units except in the case of impact from above. Basin water serves as an additional confinement barrier and heat sink.

Resources

Approximately 22,000 tons of carbon steel and 45,000 tons of stainless steel will be used through the year 2010. Concrete usage will be about 140,000 cubic yards (110,000 m^3). Land requirements will be about 150 acres (60 ha). Personnel requirements will be 95 in 1983, 330 in 2010, and 160 thereafter. Water usage will be about 1,500 million gallons (1.7 million m^3) per year in 2010, and electrical usage 10,000 kilowatts.

Technology

The handling and storage of radioactive materials under water is a standard method of operation in the nuclear industry. For over 25 years, reactor fuel elements have been routinely stored in water basins. Water basins are also used for handling radioactive sources at research facilities. However, the storage of high-level waste canisters with stainless steel overpacks in water at 120 °F (50 °C) for 100 years is an extension of present technology. Care in the design and operation of systems for chemical adjustment and control of basin-water quality will be required

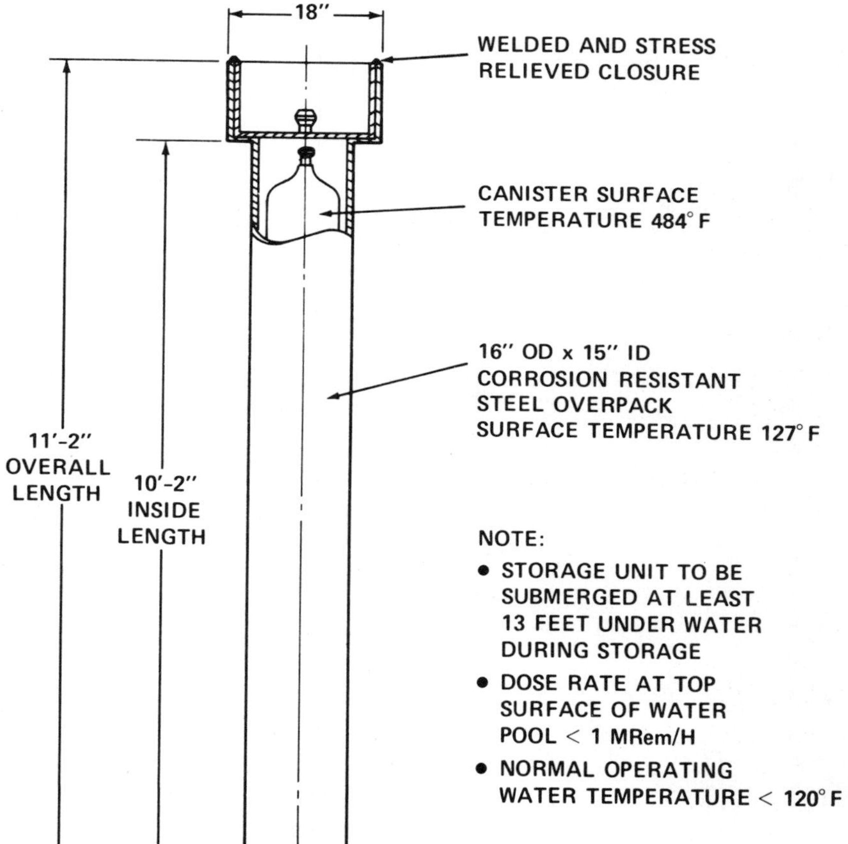

FIGURE 8
Diagram of a preliminary AEC water basin concept—overpack storage unit.

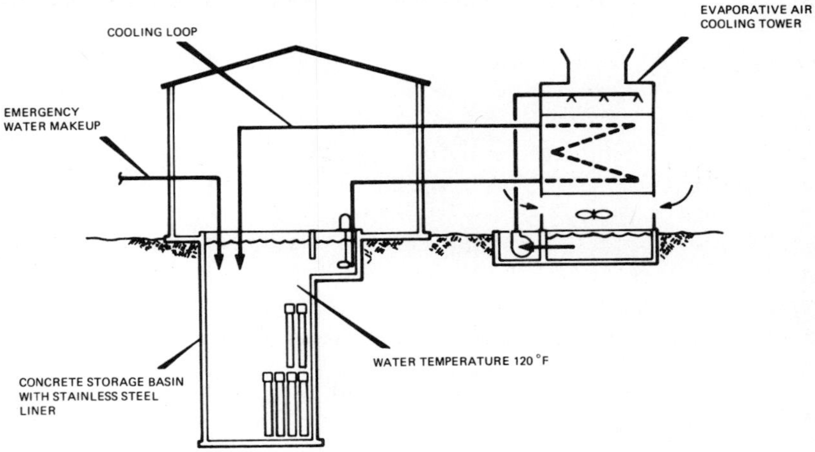

FIGURE 9
Diagram of a preliminary AEC water basin concept—overpack facility.

to minimize corrosion during storage. Full protection from the effects of severe earthquakes, tornadoes, sabotage, and aircraft impact, is much more difficult than for the SSCC.

Summary Evaluation

Of all the concepts being considered, the greatest operating experience exists for the water basin concept. However, the Panel criterion of passive cooling is not met because a pump is required to circulate the basin water through a heat exchanger. In addition, the criteria for container integrity and contamination-free water require rigorous overpack designs. Canisters and overpacks must be essentially stress-free; also, they must be able to withstand exterior attack by coolant impurities and interior attack by components of the waste.

AIR-COOLED VAULT CONCEPT

In the air-cooled vault concept, each waste canister is placed within a 0.5-inch (1.3-cm)-thick, low-carbon steel overpack with a 16-inch (40-cm) outside diameter and sealed with a welded plug (Figure 10). The overpacks are suspended vertically in concrete vaults, each holding 500 overpacks, surrounded by steel sleeves (Figure 11). Cooling air, filtered by a coarse screen, enters through side inlets and a lower distribution plenum, and passes upward through annuli between each overpack and its sleeve. Heated air from the annuli is collected in an upper plenum and is then discharged to the atmosphere, unfiltered, through a slab-shaped vertical stack 21 feet high, 4 feet wide, and 60 feet long (6.4 m X 1.2 m X 18.3 m). Circulating air will maintain an overpack surface temperature of 400 °F (205 °C) and a canister surface temperature of 620 °F (325 °C); the air will be heated to about 210 °F (100 °C). Air circulation is by natural convection up the 21-foot (6.4-m)-high stack. Maximum air speed up the stack is 6 feet (1.8 m) per second.

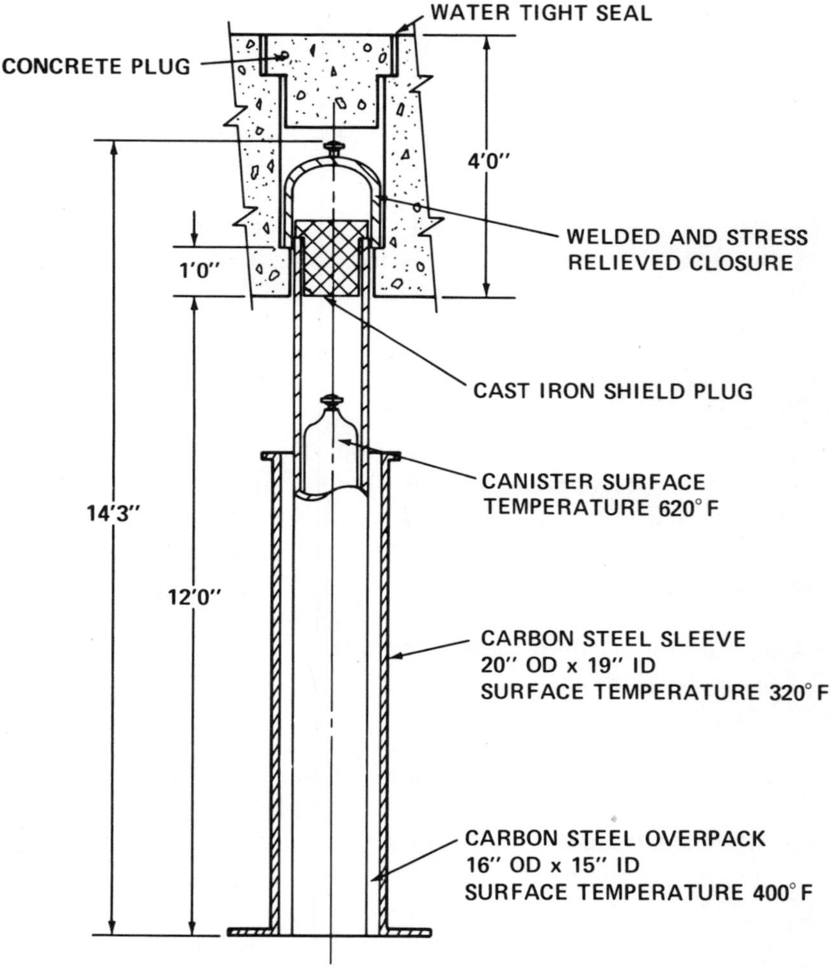

FIGURE 10
Diagram of a preliminary AEC air-cooled vault concept storage unit in cell.

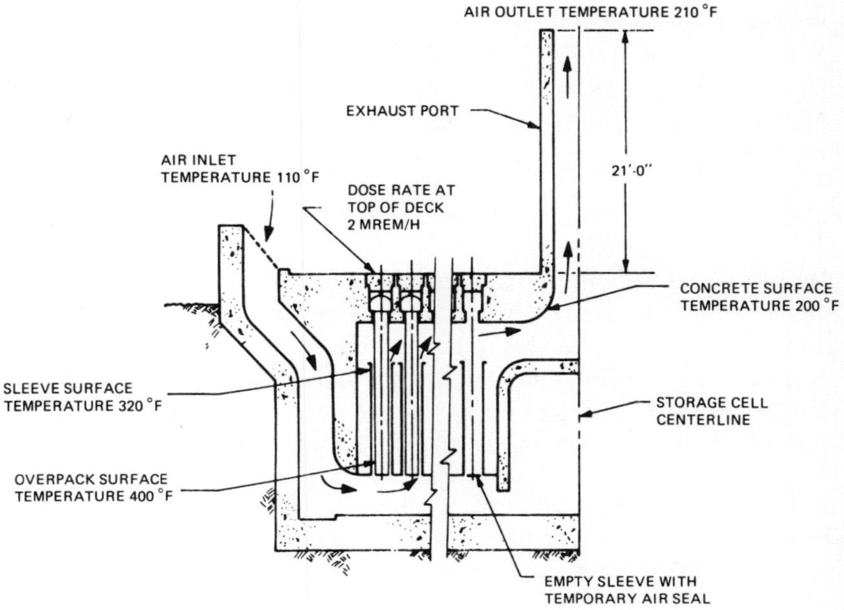

FIGURE 11
Diagram of a preliminary AEC air-cooled vault concept storage cell (500 units per cell).

Salient Features

Radiation shielding is provided by the side walls and by the 4-foot (1.2-m)-thick concrete roof of the vault and by a cast-iron plug placed above each waste canister inside its overpack before the overpack closure is welded in place. The radiation level on top of the roof of the vault will be less than 2 mrem/h.

Storage units would be constructed in groups of seven vaults, holding a total of 3,500 canisters. To accommodate the 75,000 canisters expected by the year 2010, 22 such storage units would be required.

The ACVC involves a passive cooling system and high-integrity overpacks physically protected by the concrete vault structure. Absence of filtration of the effluent air places total reliance on continued integrity of the overpacks. Continual surveillance to ensure that air passages remain unobstructed is required.

Resources

Approximately 150,000 tons of steel, of which 75,000 tons will be for overpacks, and 340,000 cubic yards (260,000 m^3) of concrete will be used through the year 2010 to store the expected 75,000 canisters. Personnel requirements will be approximately 80 in 1983, 280 in 2010, and 90 thereafter. In 2010, water usage will be about 74 million gallons (280,000 m^3) per year and electrical usage 2,100 kilowatts. The total area required to store 75,000 canisters is about 100 acres (40 ha).

Technology

No further research or development on the principles of heat transfer and air flow, by which heat would be removed from the wastes, is required; however, there is a question as to whether the proposed slab-shaped stack would assure a sufficiently reliable air flow to maintain temperatures below design limits. A duct 4 feet (1.2 m) wide, 60 feet (18.3 m) long, and only 21 feet (6.4 m) high is possibly subject to Taylor instabilities (in which local down-drafts of cool air reduce the effective draft and the effective air-circulation rate). Model experiments should be made to determine whether design air-flow rates can be maintained through the cooling annuli, both with fully loaded and with partially loaded vaults, over the full range of wind speeds and directions.

Tests to establish that a 0.5-inch (1.3-cm)-thick carbon steel overpack will remain leak-tight for 100 years under the expected conditions of temperature and radiation are imperative because the overpack is the last barrier against widespread dispersal of radioactive material, if the canister should leak. This natural circulation system cannot incorporate an effective filter system in the discharge air ducts.

Summary Evaluation

The most attractive features of the ACVC are its predominantly passive character, low capital cost, and small requirement for land. This concept appears to be less certain to prevent dispersal of radioactive materials because of uncertainty about the 0.5-inch (1.3-cm)-thick carbon steel overpack that surrounds each canister. It may be difficult to demonstrate in advance that this thickness of steel is sufficient to protect against leaks for 100 years, especially in the presence of an intense radiation field. If the overpack fails, the relatively high air speeds through the annuli and stack might disperse radioactive materials to a greater extent than with the other concepts. Finally, the slab-shaped stack of this concept is vulnerable to damage by earthquake, tornado, sabotage, or aircraft accidents.

ENVIRONMENTAL CONSIDERATIONS

The well-being of the people of the United States rests on an adequate and readily available supply of energy, for which nuclear fission is an important prospective source. Producing nuclear power involves certain environmental and health-related hazards. These hazards have probably received more detailed study than those for any other power-producing system, and their containment to reasonable limits appears to be well within present scientific and technological capabilities. The ecology, hydrology, geology, and climate are important considerations in the selection of site for a retrievable surface-storage facility.

ECOLOGY

As with all environmental issues, even those less complex than the storing of high-level radioactive wastes, there is no single or simple path to a "best" solution. There are, instead, a number of pathways to a variety of solutions.

Complex interrelationships among land, air, and water resources and interchange of pollutants among these resources and biological systems are factors to be taken into account in environmental-quality analysis and regulation. Unfortunately, there are no universally accepted quantitative measures of health, safety, ecological stability, or aesthetic values.

A detailed environmental-impact analysis is required by federal law in order to develop a complete and valid comparison of deleterious environmental impacts and social benefits to be derived. An ecological inventory must precede the environmental-impact analysis.

Ecological impacts of the several schemes for interim storage of solidified high-level radioactive wastes have four major sources: (1) facility construction; (2) transportation of the wastes; (3) release of radioactivity; and (4) release of heat into the storage-site environment.

Construction will disturb local ecosystems, in part by displacing or destroying indigenous species. A storage site will restrict migration of plant and animal populations by its surrounding security structures and its interior roads and other compacted surfaces. The site (or sites) preferably would be located where natural species are few and individuals not abundant, and where an extensive buffer zone is available to absorb displaced populations.

Safe and reliable transportation of the wastes is required for each interim storage concept. As rail transport is most adaptable to the large weights involved, the existing network of rail lines becomes a principal factor in site selection. During transport, canisters of solidified waste, which have been proof-tested against rupture in the event of an accident, will be enclosed in shipping casks designed to maintain an external temperature and a radiation rate at acceptable levels. Thus, the hazard of transporting waste can and must be held at or below that level that results from transporting other potentially hazardous commodities over the same route. (See Appendix H.)

Other sections of this report describe the multiple barriers against dispersal of radioactive materials and anticipated intensities of radiation within and at the boundary of a storage facility. These design objectives must consider the effects of (1) possible concentration of initially inconsequential amounts of radionuclides or toxins in the tissue of plants or animals or in the soil by physical and biologic processes; and (2) the mobility of rodents, reptiles, small mammals, insects, and birds.

HYDROLOGY

In broadest terms, the requirements for an acceptable retrievable surface-storage facility site would, among other essentials, include (1) perennially unsaturated material beneath the foundations of principal structures, to ensure the permanent stability of those structures; (2) a dependable and adequate local source of water of acceptable quality; and (3) effective management of both routine effluents from the site (radioactive or not) and occasional accidental discharges of radioactive products. Particular criteria would vary according to the scheme of storage and design of the structures.

In general, the site should be above the highest conceivable flood level of any adjacent perennial stream and above any water level that conceivably might result from failure of any up-valley dams or from down-valley landslides. The site should also be beyond the reach of, or readily protectable from, "flash" runoff in intermittent or ephemeral stream channels. Further, beneath the site, depth to the water table should be greater than depth of foundations to assure that foundation materials remain unsaturated. Thus, under the designs suggested in *Retrievable Surface Storage Facility: Alternative Concepts, Engineering Studies,*[17] depth to the water table would have to be about 20 feet (6 m) or more for a sealed-cask facility, 30 feet (9 m) or more for an air-cooled vault facility, and 50 feet (15 m) or more for a water basin facility. Still greater depths to the water table might be desired for managing effluents, as will be outlined.

The water supply required by a retrievable surface-storage facility during early years of operation would be only moderate, whatever the type of that facility—for a sealed-cask or air-cooled vault facility, about 30 million gallons (0.113 million m^3) a year, [60 gallons (0.23 m^3) per minute on the average]; and, for a water basin facility, slightly more than twice as much. The ultimate requirement would be about 2½-fold greater for the sealed-cask and air-cooled vault facilities, but 25-fold greater for a water basin facility. In other words, the ultimate requirement of a water basin facility would be about 1,500 million gallons (5.6 million m^3) a year [or nearly 3,000 gallons (11.2 m^3) per minute]; the requirement during an emergency (such as a ruptured water basin) could be considerably greater. (See ARH-2888 Rev.)

Ordinarily, a facility site would be remote from an existing developed water supply, and an adequate local source would therefore be essential. For the water basin concept, a standby source of several days' supply would be necessary for an emergency. The smaller requirement of a sealed-cask or air-cooled vault facility could be satisfied from a groundwater source of average or greater yield. The relatively large ultimate requirement of a water basin facility could be satisfied only from a perennial stream or lake or from a groundwater body of uncommonly large yield. In much of the United States, especially in its semi-arid western part, use of a stream or groundwater body is regulated under the doctrine of prior appropriation, by permit, or under other legal constraint. Hence, the water supply for a prospective retrievable surface-storage facility would be competitive, under minimal priority, with all prior supplies for other purposes. Thus, for a water basin facility, the ultimate consumptive water requirement—about 50 percent of the withdrawal requirement[17]—could be difficult, or even impossible, to satisfy except by purchase of an antecedent water right. The earliest appraisal of any prospective site for a retrievable surface-storage facility should determine the location, potential perennial yield, and legal availability of local water sources.

Chemical quality of the water supply is likely to be critical only in the case of a water basin facility. For such a facility, all the basin water would have to meet dissolved solids standards equal or superior to those of a public water supply and have a chloride content preferably no more than 0.5 part per million.[17] These requirements could, of course, be met by demineralizing the native raw water—at a cost and under the necessity of effectively disposing of wastes from the demineralizer. Thus, water-quality limitations conceivably might preclude an otherwise qualified ocean-front site or an inland site in an area of brackish native waters.

Surface drainage of a site should be readily controllable to the end that no runoff of rain, snow melt, or process water conceivably could dissolve or entrain radioactive substances and then escape the site without passing a suitable monitoring station.

All site-generated liquid waste, both nonradioactive and radioactive, must be readily manageable by such treatment as may be suitable and advisable, followed by return to a stream or percolation into the ground. In the latter instance, adsorption and base exchange in earth materials both above and below the water table, as well as the pattern of circulation and rate of movement below the water table, all must be predictable. To this end, investigations for selection of a site must establish depth and configuration of the water table, also transmissivities and adsorptive or base-exchange capacities of all strata that might be invaded by percolating wastes.

The limitations on site selection are greatest for the water basin concept because, in this case, there must be an assured, permanent supply of water of adequate quantity and quality, while at the same time all essential criteria required for the other concepts must be met. Such a requirement eliminates many potential sites in semi-arid or arid regions that otherwise would meet the siting criteria.

GEOLOGY

In addition to hydrogeologic criteria, several additional aspects of the physical environment must be considered. For the concepts of water basin or air-cooled vault, and to a lesser degree for the sealed casks, the site must be of low seismicity.

In all concepts, as a protection against accidental spills, the surficial materials should have both low permeability and substantial ion-exchange capacity. This is of particular significance in the water basin concept because, if a rupture of the basin produced leakage, a ground material with low permeability and a high exchange capacity could contain the spill and prevent the dispersal, or slow its movement downward to the water table. The topography, geology, and climate must be such as to ensure that mud flows or catastrophic landslides will not reach the area, with the possibility of disrupting or burying the storage facility.

CLIMATE

Local climate has a bearing on all waste storage concepts. Clearly, a cold climate that has resulted in permafrost would not be acceptable for any of the concepts, but a cool climate is desirable to assure that the designed operating temperatures of a sealed cask or of an air-cooled vault facility would be realized. In this connection, design criteria for a sealed cask facility assume an ambient air temperature of 80 °F (28 °C). The assumed ambient air temperature for an air-cooled vault is 110 °F (43 °C).

Whatever its type, a waste storage facility on the land surface will discharge a considerable quantity of heat into the adjacent atmosphere (also, in the case of a water basin facility, a considerable quantity of water vapor). Microclimate at and near the site will be influenced inevitably. Possible hazardous effects include aggravated fogging and icing; potentials for such effects should be evaluated in the preliminary examination of each prospective site.

Local humidity will influence the life expectancy of air-cooled, metallic waste containers, e.g., canisters, overpacks, sleeves, or casks. Here, a relatively dry climate would tend to minimize corrosion but might not be required if suitable materials were chosen for fabricating the containers. Also in this connection, the microclimate of an industrial area might be adverse owing to airborne agents of aggravated corrosion, including halides or oxides of sulfur. A dry climate would also minimize or eliminate the possibility of local ponding or sheet flooding at the site.

COST ESTIMATES

The available cost estimates of the various concepts are preliminary, but they do provide insight into the economics of a retrievable surface-storage facility (RSSF). The following discussion is directed to establishing—

1. the magnitude of the costs involved in RSSF and the effect of these costs on power generation;
2. major differences among the concepts; and
3. the necessary initial capital commitments *vs.* costs that may be deferred to the time of use, i.e., financing flexibility.

Table 5 compares the estimated costs for the three concepts. There are two major categories of costs, capital and operating. The capital costs include all fixed facilities, the overpacks in the case of the air-cooled vault concept (ACVC), and the casks and shields in the case of the sealed storage-cask concept (SSCC). The operating costs are the recurring expenses of staffing, maintenance services, and utilities. The following major assumptions underlie these estimated costs:

1. All wastes are received at a RSSF within 10 years of generation.
2. The initial RSSF will be ready to operate in 1981, receive its first

wastes in 1983, and expand as needed to fulfill waste storage requirements through 2010 (75,000 standard canisters).

3. The RSSF is AEC-owned and contractor-operated.
4. Hanford is the "reference site" for site-dependent cost estimates.

The capital cost estimates developed by Kaiser Engineers for AEC are based largely on a preliminary engineering study, but include contingencies to allow for the limited work on the various concepts. Operating costs were similarly estimated by the Atlantic Richfield Hanford Company (ARHCO).

Although the estimated costs are approximations, the noteworthy observations about Table 5 include the following:

1. The initial capital investments (Item A-1, Table 5) are approximately the same for all three concepts. This is readily explained by the fact that the initial investment is dominated by the costs of the receiving and handling facilities, which are similar in all concepts.
2. The total capital outlay for the SSCC is greater than for the other two concepts, but the added increment represents the cost of casks and shields, which need not be budgeted in advance of requirements for added storage capacity. However, while all three concepts require

TABLE 5

Comparison of Cost Estimates (1973 dollars)

	Sealed Storage-Cask Concept	Air-Cooled Vault Concept	Water Basin Concept–With Overpack
A. Capital ($ million)			
1. Initial facility (1981)	20	25	25
2. Added facilities (thru 2010)	60	280	315
3. Casks, overpacks, and shields (thru 2010)	820	155	400
TOTAL CAPITAL	900	460	740
B. Operating costs ($ million)[a]			
1. Thru 2010	200	165	220
2. Thru 2110	660	845	1,400
C. Totals (thru 2110)			
1. In $ million	1,560	1,305	2,140
2. In $ thousand/waste canister	20.8	17.4	29.6
3. Contribution to electrical energy cost (mill/kWh)[b]	0.026	0.022	0.036

[a]Includes all costs except fixed capital, overpacks, casks, and shields.
[b]Each canister of waste results from the generation of approximately 800 × 10^6 kWh of electricity.

the same type of major handling facilities, the cost for Items A-1 and A-2 in the case of SSCC facilities is about one-fourth that for the other concepts.

3. The operating cost estimates through 2010 (Item B-1), the period during which waste is being delivered and prepared for storage, are comparable for all three concepts. This is again due to the fact that the major operating requirements during this period are similar for all concepts. The total operating cost estimates through 2110 are dominated by the expenditures during the 100-year surveillance period. Here we note a significant cost advantage for the natural circulation air-cooled concepts (SSCC and ACVC), which eliminate the costs of water and power required by the WBC and reduce staffing requirements.

4. The total cost estimates to own and operate a retrievable surface-storage facility, including surveillance for 100 years after the last waste delivery, are similar for the two air-cooled concepts, but about 50-percent greater for the water basin concept. On the basis of the assumed ultimate capacity of 75,000 canisters, the cost estimates per canister are between $17,000 and $30,000. The contribution to electrical energy costs by any of the interim storage concepts considered is estimated at only a few hundredths of a mill/kWh. This compares with 10 to 15 mill/kWh for the total electrical generating cost for nuclear power plants.

RELIABILITY OF MATERIALS

The satisfactory performance of a retrievable surface-storage facility depends upon the reliability of the materials of construction. Assuring such performance is a challenge because the 100-year designed lifetime exceeds, by a factor of two to five, that which usually is projected for most industrial works.

Available engineering materials and technology are capable of providing a materials system that is adequate to perform satisfactorily for at least 100 years, provided that unexpected external forces do not occur that exceed the reasonable design capacity of the RSSF. Many metal bridges and other civil engineering structures have been in continuous use for more than 100 years. Each concept provides for the overpacking of a canister as needed.

In some specific aspects, the optimum route for achieving a reliable and reasonably economical waste storage facility is not clear. For example, although the corrosion rates of many alloys are virtually negligible in a wide variety of environments, certain abnormal circumstances greatly accelerate such rates even at room temperatures. In these circumstances, metals may perforate at low stress. Thus, design for satisfactory performance of materials depends upon the assurance with which the chemical and physical stability of the waste materials and their external environments can be foreseen and controlled. Uncertainty in this regard requires a commensurately more conservative design.

In the largely unproven field of managing radioactive wastes over the long term, it becomes crucial that (1) specifications for all construction materials and for processes be proven by rigorous prior testing and by monitoring during operation of any RSSF; (2) designs for components of the facility and processes be adjusted as shown by accumulating data to be desirable and necessary; but (3) no changes in process, materials, or design be permitted before critical tests have been conducted.

This proof testing should aim at predicting performance under steady-state and variable, normal and abnormal (accident) conditions. Further, criteria for extrapolating long-term performance from short-term tests should be developed and evaluated. The crucial proof tests just outlined should be reviewed and confirmed at each of their stages.

The following subsections emphasize particular aspects of the reliability of materials in the three retrievable storage concepts examined by this report. They are based on conservative standards with regard to forces and influences that must be withstood in normal operation of a storage facility, including accidents in handling and transporting radioactive materials. However, extraordinary external forces, such as aircraft impact, explosion, sabotage, and extreme natural disasters are not considered. A summary of potential abnormal conditions and maximum credible accidents is given in WASH-1539, Section 3.[18]

COMPOSITION AND FORM OF THE WASTES

The first of the barriers in preventing contamination of the environment is the form of the wastes. Special attention should be given to optimizing the thermal, chemical, and mechanical stability of the wastes and their expected lives.

It is understood that the high-level wastes to be stored (1) derive from the reprocessing of fuel from commercial nuclear power reactors of the present light-water type; (2) per current AEC regulations, have been reduced to a solid which is chemically, thermally, and radiologically stable; and (3) have been aged at least 5, but not more than 10, years. These presumptions place only very general limits on chemical composition and physical form of the wastes. It is anticipated that composition will vary moderately and that form may change if processes for packaging the wastes are modified.

For assurance that a particular retrievable surface-storage facility will be designed to perform reliably, the following two specific criteria for stability of the wastes must be known to be satisfied whether the surrounding environment is gaseous or liquid, however intense the radiation field (gamma and neutron), and however great the temperature:

1. The wastes shall not release species that can chemically attack the metals of canisters, overpacks, or casks—especially species that are potential agents of stress-corrosion cracking. In particular, such release shall not be by separation of a molten phase.
2. The wastes shall not release gases increasing internal pressure within canisters.

The objective of these criteria is that the nature of a waste material shall not lessen the mechanical integrity of a canister in which that waste is contained. As has been implied, the criteria must be satisfied, not only under normal conditions of storage facility operation, but also under the conceivable range of abnormal conditions (accidents). Thermal abnormalities due to prolonged lack of cooling or—during transportation—to slumping of waste that is in a granular form, are included.

As stated, the early prototype form of solidified waste has been a fine granular calcine. Compacted oxide or other ceramics are superior forms, not only to satisfy the preceding criteria, but also because such optimized stable, compact, insoluble ceramics are themselves the innermost barrier against dispersal. Recommendation 5a of this report proposes to devise an optimum form.

CANISTERS

Canister integrity is the second barrier to the release of radioactivity. Principal factors are (1) the canister material and manner of fabrication, and (2) the external and internal environments to which the canister will be exposed.

The canister should meet the following standards:

1. The material should have minimum susceptibility to corrosion, general or localized, in any anticipated environments, external and internal. In this regard, austenitic stainless steel may not be the most advantageous material because, although it has excellent resistance to general atmospheric corrosion and to many chemical agents, it is very susceptible to localized corrosion in environments that contain chloride, sulfide, or certain other agents. Careful attention should be given to selecting the optimum alternative material.

2. Residual stress in a fabricated and sealed canister should be minimized to reduce the potential for accelerated corrosion of stressed areas.

3. Following fabrication, no contaminants with the potential to engender localized corrosion, either by themselves or by reaction with the environment, such as welding flux or machining lubricant, should remain on canister surfaces.

The external environment of a canister ordinarily would be gaseous (air, with its local content of water vapor and contaminants) for both the sealed-cask and the air-cooled vault facilities. It would be aqueous for a water basin facility if an overpack should leak. In this circumstance, design for corrosion-free performance should avoid crevice-like spacing of metal parts and of unlike metals near one another that might react galvanically (for example, as among canister, overpack, and cask). The facility would be sited preferably in an area of small humidity remote from industrial plants from which agents of corrosion might be airborne in abundance. In the case of a water basin facility, demineralizing the native water might be necessary to minimize potential agents of corrosion, such as the chloride ion. (This requirement has been noted previously.)

The external environment would also be aqueous while a filled canister was held at a fuel-reprocessing plant before being transported to the prospective storage facility. During this interim, a canister probably is more sensitive to abusive exposure than at any other time; monitoring of its performance would be essential.

The internal environment of canisters ordinarily would be the same for all three RSSF concepts—gaseous—owing to residual air entrapped when a canister was sealed or a ruptured canister was overpacked. In the gaseous environment, assuming that preceding criteria for stability of waste materials are satisfied, the potential for internal corrosion should be very small. An aqueous internal environment, as from rupture of a canister in a water basin facility, would be analogous to conditions already discussed.

OVERPACKS AND CASKS

The overpack or cask provides the third barrier. In general, the overpacks or casks would be exposed more to external corrosion than canisters would be. Also, because they are bulkier and heavier, they are perhaps more susceptible to mechanical damage, as by being dropped accidentally. Thus, selection of materials and design should consider the following—again, over the anticipated and reasonably conceivable ranges of temperature, humidity, and both airborne and water-borne agents of corrosion:

1. The material, steel in general, should have high fracture resistance.
2. Walls should be no thicker than required for functional reliability, to the ends of easy inspection and handling, minimum residual stresses, and a near-planar state of stress.
3. Crevice-like spaces can retain water and so accelerate localized corrosion, also induce localized stress due to expansion of corrosion products. Obviously, such spaces are to be avoided.
4. Galvanic attack should be avoided by separating dissimilar metals or by cathodic protection, as relevant.
5. Materials should have minimal susceptibility to stress-corrosion cracking, pitting, and intergranular attack. Attention should be given also to potential for the leaching of corrosive agents from surroundings—for example, chloride (from concrete), sulfide, or others.
6. Permissible residual stress should be no greater than that specified previously for canister materials.

CONCRETE SHIELDS

The concrete shield constitutes the fourth barrier. The structural concrete used in the proposed storage concepts is, in general, subjected to low stresses under normal operating conditions. For example, in the shield for the 5-kilowatt SSCC the thermal stresses would be about 500 psi tension on the outer surface and 1,100 psi compression on the inner surface. In the other concepts thermal stresses would be even less, and normal design practices would ensure that other stresses would be within allowable limits. There should thus be no difficulty in providing concrete of adequate strength and durability, provided the concrete does not deteriorate substantially with time.

The sealed storage-cask concept requires the closest attention to possible limitations of structural concrete. Normal operating conditions can be satisfied easily, even though the thermal gradient will produce cracking on the outer surface of the shield. Closely spaced reinforcing bars will control crack size and preserve the integrity of the shield. Aspects of concrete behavior that need further attention include diminution of concrete strength and durability as functions of time and temperature, thermal stresses under abnormal operating conditions, and possible loss of strength under radiation.

The concrete shield for the 5-kilowatt SSCC will operate at 155 °F

(70 °C) inside and 84 °F (30 °C) outside temperatures, with peak temperature for zero air flow reaching 650 °F (345 °C). These and other expected temperatures in the concrete components of those concepts that use structural concrete follow:

Concept	Normal	Abnormal
WBC	120 °F (50 °C)	210 °F[a] (100 °C)
WBC-OP	120 °F (50 °C)	210 °F[a] (100 °C)
ACVC	200 °F (93 °C)	
5 kW SSCC	84–155 °F (30–70 °C)	120–650 °F[b] (50–345 °C)
10 kW SSCC	95–255 °F (35–125 °C)	135–1190 °F[b] (60–645 °C)

[a]basin water boiling
[b]zero air flow

Considerable research has been conducted on thermal effects in concrete.[19–20] While results are commonly somewhat contradictory because the many variables are not treated alike by the several investigators, definite trends have emerged. Thermal effects pertinent to this report may affect—

1. strength, compressive and tensile;
2. durability;
3. shielding characteristics; and
4. thermal properties.

Thermally induced changes in strength of concrete vary considerably with such factors as type of aggregate, curing conditions, temperature range, degree of thermal cycling, and the surface conditions of the concrete, i.e., sealed or unsealed. The last factor is important because a sealed surface inhibits the natural release of water as temperature increases, and so leads to increased thermally induced stresses and cracking.

Prolonged temperatures of about 200 °F (93 °C) acting on a limestone-aggregate concrete, will reduce the compressive strength of the concrete by as much as 40 percent, primarily because of the differences in coefficients of thermal expansion of the aggregate and of the cement matrix. Aggregates with greater thermal coefficients (basalt, medium-silica-content crushed rock, etc.) will substantially diminish this loss in compressive strength. With the proper choice of aggregate and under normal operating temperatures, the shielded 5-kilowatt SSCC will not lead to consequential changes in concrete strength.

Thermal cycling is usually more damaging to concrete than is a steady-state thermal environment, although this may not be true for thoroughly dried concrete. While temperature within the concrete would vary both diurnally and seasonally, these variations would be small compared with those in the ambient air. However, the effect of one or more severe thermal cycles (no-air-flow condition) over the intended life of the shield is unknown and should be appraised.

Tensile strength of concrete is more sensitive to temperature than is compressive strength. Adequate reinforcing steel can lessen tensile cracking and withstand mechanical and thermal stresses; however, any tensile cracking can diminish the conduction of heat, and possibly the biological shielding effectiveness of the concrete as well.

Finally, greater temperatures in concrete tend to lessen the attenuation of any neutron flux because of the lower content of water in the hotter concrete. For the normal operating conditions of the SSCC, this behavior is not a serious disadvantage and can be compensated for in design.

For the no-air-flow condition, a temperature drop of 530 °F (277 °C) across the concrete shield of the 5-kilowatt SSCC is anticipated. Given such a drop, linear-elastic stress theory (with $E = 3.5 \times 10^6$ psi) predicts outer-surface tension of 3,200 psi and inner-surface compression of 6,700 psi. Since the concrete can withstand only a few hundred psi tension, the predicted tension stress would cause extensive cracking and a gross redistribution of stresses in the shield. As has been stated, crack size can be minimized by proper steel reinforcing, but the effects of this cracking on the thermal properties of the shield are uncertain. Also, the cracking would seriously lessen the shielding characteristics of the concrete.

The inner-surface stress of 9,400 psi biaxial compression is also cause for concern. This stress state, and the anticipated extreme temperature of 650 °F (343 °C) would produce compressive failure in even high-strength concrete. High-strength concrete, such as that used in prestressed concrete construction, would not be acceptable for use in the shields, unless provision is made for their replacement when necessary. Further, the thermal-stress state would produce a tendency for tensile cracking (Poisson-induced strain) normal to a cylindrical path near the inner surface. Under the postulated abnormal operating conditions, rather substantial axial and hoop reinforcing steel would be needed to minimize the effects of cracking, crushing, and spalling along the interior surface.

A fully nonlinear, finite-element, thermal-stress analysis of a simulated full-scale reinforced-concrete shield—in which the concrete has cracked and been crushed and the steel may have yielded under the extreme operating condition—is needed to assess the true stress state and thermal gradient in the shield. Actual stresses would be substantially less than the predicted linear-elastic stresses because modulus E decreases as temperature increases, and because stresses are redistributed after tensile cracking occurs. On the other hand, the coefficient of expansion of concrete tends to increase with increasing temperature, which would lead to greater thermal stresses. A combined experimental and analytical program is necessary to verify the design and the projected thermal and shielding properties of the concrete shield for a 5-kilowatt SSCC.

A 10-kilowatt SSCC shield, with abnormal temperatures estimated to be 1190 °F (643 °C) inside and 135 °F (57 °C) outside, would place intolerable structural demands on the concrete. Reinforcing steel could not prevent inner-surface failure because the steel would lose most of its stiffness and tensile strength at the high interior temperature. The distressed concrete would lose much of its ordinary thermal conductivity so that temperatures even higher than those estimated might ensue. Thus it is believed that the shields for 10-kilowatt casks would fail in a no-air-flow incident. If such failure should occur, however, it would be relatively easy to provide a new concrete shield.

The design values of neutron flux range from 1.4×10^4 to 1.1×10^7 neutrons/s cm^2. Test results on typical low-stress concrete at ordinary ambient-air temperatures have shown that this range of irradiation should not produce any significant change in concrete properties. However, SSCC operating conditions invoke the inverse and much more severe question—that is, what are the effects of greater-than-ordinary temperature and thermal stresses on the shielding characteristics of concrete?

Since concrete properties vary so widely as a function of the aggregate, it is imperative that the actual composition of cement and mix of the concrete to be used in the shield be determined and defined explicitly. Factors to be considered include—

1. change in thermal properties as a function of the temperature–time history, and the effect of these changes on calculated thermal environment in the shield;
2. effect on the heat transfer and shielding characteristics of cracking, which may result from thermal gradient or moisture gradient abnormal operating conditions; and
3. long-term effect of temperature and radiation on the properties of concrete.

An extensive experimental program is needed to resolve these questions.

REFERENCES

1. U.S. Atomic Energy Commission, *Management of Commercial High-Level and Transuranium–Contaminated Radioactive Waste,* AEC environmental statement, WASH-1539 (Sept. 1974). (draft)
2. *Code of Federal Regulations,* Title 10 CFR, Part 50, Appendix F (10 CFR 50, Appendix F).
3. Radiation to the general public shall be based on the quantitative guidelines of the proposed Appendix I to 10 CFR 50 (less than 5 mrem per year). For areas where the public has limited access, the design objective is to limit radiation exposure to less than 10 percent of that allowed in AEC Manual 0524, Standards for Radiation Protection.
4. Heroy, W. B., *Disposal of Radioactive Wastes in Salt Cavities*, Report prepared for Committee on Disposal of Radioactive Waste Products in Geologic Sub-structures by the Sub-committee on Deep Disposal, NRC Division of Earth Sciences, National Academy of Sciences, Washington, D.C. (Feb. 1957).
5. Muehlberger, W. R., and E. F. Gloyna, *Geology of Salt Domes and Storage Possibilities*, Technical Report to U.S. Atomic Energy Commission, The University of Texas Sanitary Engineering Research Lab, Austin (Nov. 1958).
6. Committee on Waste Disposal, *Disposal of Radioactive Wastes on Land*, Division of Earth Sciences, NAS-NRC Publication 519, National Academy of Sciences, Washington, D.C. (1957).
7. Culler, F. L., J. O. Blomeke, and W. G. Belter, *Current Developments in Long-Term Radioactive Waste Management*, Vol. 2, Fourth International Radioactive Waste Conference, Geneva, Proc., (Sept. 1971).
8. Lomenick, T. F., and A. L. Boch, *Site Investigation for a Bedded-Salt Pilot Plant in Permian Basin*, Vol. 1, Second International Symposium on Underground Waste Management and Artificial Recharge (Sept. 1973).

9. Panel on Disposal in Salt Mines, Committee on Radioactive Waste Management, *Disposal of Solid Radioactive Wastes in Bedded Salt Deposits*, National Academy of Sciences–National Research Council, Washington, D.C. (Nov. 1970).
10. Pierce, W. G., and E. I. Rich, *Summary of Rock Salt Deposits in United States as Possible Storage Sites for Radioactive Waste Materials*, U.S. Geol. Survey, Bull. 1148 (1962).
11. Schneider, K. J., and A. M. Platt, *High-Level Radioactive Waste Management Alternatives*, BNWL-1900 (4 vol.), Report to U.S. Atomic Energy Commission, Battelle Pacific Northwest Laboratories, Richland, Wash. (May 1974). [Summary issued as WASH-1297.]
12. Kuhn, K., *Salt Mine, Federal Republic of Germany–Operating Facility for Underground Disposal of Radioactive Wastes,* Vol. 2, Second International Symposium on Underground Waste Management and Artificial Recharge (Sept. 1973).
13. Clelland, D. W., *High-Level Radioactive Waste Management in the United Kingdom,* British Nuclear Fuels Limited, Risley, England, Symposium on Management of Radioactive Waste From Fuel Reprocessing Plants, Paris, November 27–December 1, 1962 (OECD Mar. 1963).
14. Girand, J. P., and G. LeBlage, *Design of an Industrial Facility for the Incorporation of Glass of Fission Products and the Storage of Highly Radioactive Glass,* Saint-Gobain Techniques Nouvallis, Courberoie, France, Symposium on Management of Radioactive Waste from Fuel Reprocessing Plants, Paris, November 27–December 1, 1962 (OECD Mar. 1963).
15. Blomeke, J. O., C. W. Lee, and J. P. Nichols, *Projections of Radioactive Wastes to be Generated by the U.S. Nuclear Power Industry,* ORNL-TM-3965, Oak Ridge National Laboratory, Oak Ridge, Tenn. (Feb. 1974).
16. *Retrievable Surface Storage Facility: Alternative Concept Engineering Studies,* ARH-2888, Rev., Atlantic Richfield Hanford Company, Kaiser Engineers (July 1974).
17. ARH-2888, Rev. (n. 16).
18. *Op. cit.* (n. 1).
19. *Concrete for Nuclear Reactors*, Vols. I, II, III, 1750 pp. ACI SP-34, American Concrete Institute, Detroit (1972).
20. Neville, A. M., *Hardened Concrete: Physical and Mechanical Aspects,* ACI Monogr. No. 6, American Concrete Institute, Iowa State Press (1971).
21. Claiborne, H. C., *Effect of Actinide Removal on the Long-Term Hazard of High-Level Waste,* ORNL-TM-4724, Oak Ridge National Laboratory, Oak Ridge, Tenn. (Jan. 1975).

APPENDIX A SEPARATION OF TRANSURANICS

The high-level radioactive wastes, whose treatment and storage is the concern of this report, contain two dissimilar classes of radioactive materials, fission products and transuranic elements. The fission products are elements of atomic number less than uranium, many of which are high heat-producing beta- and gamma-emitting radioisotopes such as strontium-90 and cesium-137. Most of the fission products have half-lives of 30 years or less. The transuranic elements of higher atomic number than uranium are formed by neutron capture in uranium. These are highly toxic alpha-emitting radioisotopes, many of which have long half-lives, such as curium-245 (9,320 years), americium-243 (7,370 years), plutonium-239 (24,400 years) and neptunium-237 (2.14 million years).

In 10-year-aged wastes such as would be committed to storage, fission products are responsible for more than 99 percent of the radioactivity and heat generation. After storage for on the order of a thousand years, the radioactivity and heat generation of the fission products will have decayed to relatively inconsequential levels, whereas that of the long-lived transuranics will be almost as noxious as when first committed to storage. The following table compares the heat-generation rate of fission

Prepared by Manson Benedict.

products and transuranic elements in wastes from one metric ton of representative light-water reactor fuel, 10 and 1,000 years after discharge from the reactor.[1]

Years after discharge	10	1,000
Watts per metric ton from		
Fission products	1,030	0.0177
Transuranics	7.06	2.27

Since the relative toxicities of these constituents of the waste are roughly proportional to their heat generation rates, the principal hazard from the wastes at and after 1,000 years will be from the transuranics. Because of the long half-life of some of these elements, wastes containing transuranics must be safeguarded for times of the order of a million years before they decay to innocuous levels.

The masses of fission products, uranium, and transuranic elements resulting from conventional reprocessing of one metric ton of typical light-water reactor fuel are as follows:[1]

	Grams	Weight Percent
Fission products	35,000	86
Uranium	4,780	12
Transuranics	980	2

This is based on a burn-up of 33,000 megawatt-days per metric ton of fuel and a high-level waste containing 0.5 percent of the uranium, 0.5 percent of the plutonium, and all of the other transuranics that remain in the spent fuel.

It would be desirable to separate such high-level wastes into a fission-product fraction, a transuranic fraction, and a uranium fraction for the following reasons:

1. The fission-product fraction would contain all the material that requires design of storage facilities to cope with heat production. This fraction would need to be safeguarded for times of the order of 1,000 years, after which cooling could be terminated and vigilance relaxed. It is believed feasible to design storage facilities for such wastes that would remain intact for times of this order.

2. The transuranic fraction would encompass only a few percent of the mass, heat-output rate and penetrating radiation of the fission-product fraction, but it would need to be safeguarded for times on the order of a million years. Consequently, quite different procedures are

desirable for safeguarding transuranic wastes compared with fission products. Methods that are being evaluated and may prove feasible include transmutation in a nuclear reactor into fissile material and then into shorter-lived fission products; transport into space outside the earth's orbit; or storage on earth in unique facilities especially designed to take into account the long life, low heat generation, and relatively low mass of the transuranic fraction.

3. The uranium fraction would need no unusual storage precautions, but should probably be handled separately from the transuranics—whatever disposal method is adopted for the latter—to reduce the cost of dealing with the transuranics.

The principal requirement for fractionation of high-level wastes is that the amount of transuranics remaining in the fission product fraction be small. Studies by Battelle Pacific Northwest Laboratories[2] suggest that removal of 99 percent of the transuranics from high-level wastes may be adequate. Studies reported in more detail by Kubo[3] indicate that removal of 99.9 percent of the uranium, plutonium, and neptunium originally present in the spent fuel, and 99 percent of the americium and curium would make the high-level wastes after 1,000 years no more than 10 times as toxic as the natural uranium ore from which the reactor fuel containing these wastes was produced.

The allowable fission-product content of the transuranic fraction depends on how this fraction is to be disposed of. If it is to be transmuted into fissile material in a light-water reactor, the permissible content of neutron-absorbing fission products, such as samarium and neodymium, must be only a few tenths of a percent of their original content in the wastes. If the transuranics are to be transmuted in a fast reactor, less complete separation of rare earths will probably suffice, but no precise statement can be made because of the lack of detailed cross-section data for transuranics and rare earths for fast neutrons. If the transuranics are to be disposed of in outer space or stored in specially designed facilities on earth, the degree of removal of fission products need not be so strict, as the controlling factors are now the contributions of fission products to the bulk and heat generation of the wastes.

A survey by Battelle Pacific Northwest Laboratories[2] indicates that fractionation of high-level wastes with the foregoing degrees of completeness is probably technically feasible, but that considerable process development work and pilot-plant testing will be required. Processes deemed most likely to be suitable are solvent extraction to remove the transuranics from the wastes with the required degree of completeness, followed by ion exchange to reduce the rare-earth content of the transuranics. Many problems to be solved in this process development include (1) dealing with plutonium and other transuranics present in the wastes as hard-to-dissolve solids; (2) selecting solvents and ion-exchange materials with the requisite selectivity and radiation stability; (3) developing analytical methods refined sufficiently to demonstrate that the residual transuranic content of the highly radioactive fission-product fraction is sufficiently low; (4) devising a sequence of process steps that will perform the required fractionation at an acceptable cost; and (5) selecting processes that will add the least additional bulk to the separated wastes.

The most feasible point at which to separate the transuranics (from the standpoint of simplifying the separation) is after the wastes have been stored as unneutralized liquid for the maximum time permitted by AEC, i.e., five years. The next most feasible point would be after the wastes have been converted to a solid calcine, which could still be readily redissolved in nitric acid. The least feasible point would be after the wastes have been converted to a borosilicate glass or some other form in which they are difficult to redissolve in nitric acid. At present, there is strong industrial sentiment in favor of converting wastes to a relatively insoluble glass soon after reprocessing in order to minimize spread of contamination if waste containers should develop leaks. The likelihood that conversion to glass will be widely practiced in a few years makes the development and implementation of technically feasible and economically acceptable processes for separating transuranics urgent, before large amounts of unfractionated wastes have been converted to glass.

REFERENCES

1. *Siting of Fuel Reprocessing Plants and Waste Management Facilities,* ORNL-4451, Oak Ridge National Laboratory, Oak Ridge, Tenn. (July 1970).
2. Bartlett, J. W. *et al., Feasibility Evaluation and R&D Program Plan for Transuranic Partitioning of High-Level Waste,* BNWL-1776, Battelle Pacific Northwest Laboratories, Richland, Wash. (November 1973).
3. Kubo, A. S., *Technology Assessment of High-Level Nuclear Waste Management,* Ph.D. Thesis, Nuclear Engineering Department, Massachusetts Institute of Technology, Boston (May 1973).

APPENDIX B
SUMMARY DESCRIPTION OF SEALED STORAGE-CASK CONCEPT—THICK WALL

STORAGE

In the sealed storage-cask concept—thick wall (SSCC—Thick Wall) for storage of high-level wastes, the canisters are placed in 14-inch-thick, low-carbon steel casks that are sealed by welding. Storage units, consisting of sealed casks and concrete neutron shields (Figure B-1), are stored outdoors vertically on concrete pads. Heat is dissipated from the casks by natural-convection air flow through the annulus between the cask and the neutron shield. The cask and canister surface temperatures will be about 170 °F and 450 °F, respectively.

The sealed casks furnish both radiation protection and confinement of waste. The thick walls of the storage casks furnish a rugged barrier for resistance to natural phenomena, such as earthquake and tornadoes. Atmospheric corrosion rates for a dry climate are estimated to permit at least a 100-year cask life, based on an average corrosion rate of 0.003 inch per year. The dose rate within the storage array is estimated at less than 4 mrem/h and at less than 1 mrem/h 90 feet from the array. The storage units are spaced in a rectangular storage array on approximately 20-foot centers for ease of operation and retrievability.

Prepared by the U.S. Atomic Energy Commission.

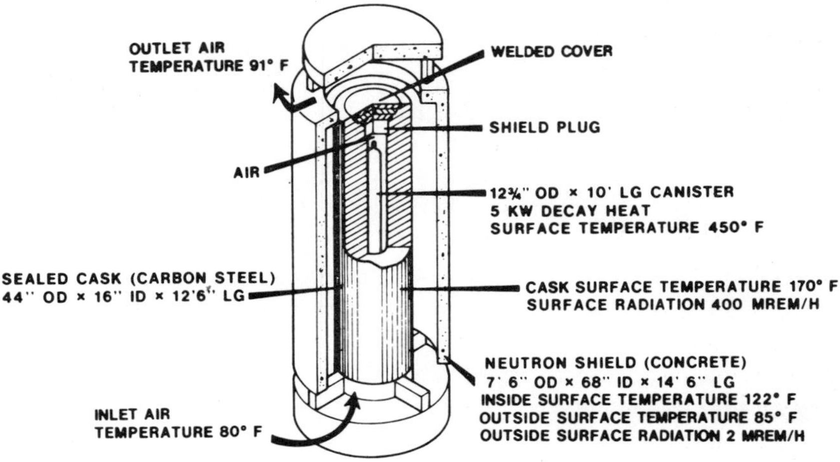

FIGURE B-1
Diagram of the preliminary AEC sealed storage-cask concept–thick wall storage unit.

OPERATING REQUIREMENTS

Land requirements will be about 1,100 acres by 2010. Personnel requirements will be about 86 in 1983, 334 in 2010, and 76 in 2011. Water usage will be about 74 million gallons per year in 2010 and electrical usage 3,000 kilowatts.

TECHNOLOGY

Heat removal from a storage unit is accomplished by a combination of radiation and convection. Air flows through the annulus between the storage cask and the concrete neutron shield by entering at the bottom of the storage unit and leaving at the top. The heat-transfer principles applied are well known.

The fabrication of heavy steel casks for at least a 100-year life has not been demonstrated. Demonstrations will be required to assure that welding and nondestructive testing operations conducted under semiremote conditions can meet necessary quality-assurance standards. Nondestructive tests will be developed to assure the long-term, leak-free integrity of a cask.

MATERIALS USAGE

Approximately 2.3 million tons of steel, of which 2.2 million tons will be used for fabrication of storage casks, will be used through 2010. Concrete usage through 2010 will be about one million cubic yards.

SAFETY CONSIDERATIONS

The SSCC–Thick Wall utilizes passive cooling and a high integrity overpack which is considered capable of withstanding postulated accidents. The concept assumes total reliance of a single thick-walled confinement barrier (overpack) to contain a waste canister. Reliable cask fabrication, testing techniques, and means to verify continued canister and/or cask integrity will require further evaluation.

ESTIMATED COSTS (1973 DOLLARS)

Total estimated program costs through 2010 are $2.8 billion, including $2.6 billion for casks and neutron shields. Operation through 2110 would cost an additional $0.66 billion.

APPENDIX C
SUMMARY DESCRIPTION OF SEALED STORAGE-CASK CONCEPT—UNSHIELDED

STORAGE

In the sealed storage-cask concept—unshielded (SSCC—Unshielded) for storage of high-level waste, the canisters are placed in low-carbon steel casks that are sealed by welding to make up storage units (Figure C-1). The cylindrical casks, with 8-inch-thick walls, are stored outdoors end-to-end horizontally in rows spaced 20 feet apart. The cask wall provides waste containment; resistance to natural phenomena, such as earthquakes and tornadoes; and partial radiation shielding. The remainder of the shielding is provided by distance. The radiation dose rate within the storage array will exceed 10 rem/h but, at the exclusion fence located 2,000 feet from the edge of the storage array to limit personnel access, is less than 1 mrem/h. All operations within the storage area must be conducted from shielded vehicles. The atmospheric corrosion rate of the cask walls will average about 0.003 inch per year.

Waste decay heat is dissipated by natural convection in the outdoor storage area. As an aid to heat transfer, each cask is supported 8 inches above-grade by two concrete support saddles. Normal cask surface tem-

Prepared by the U.S. Atomic Energy Commission.

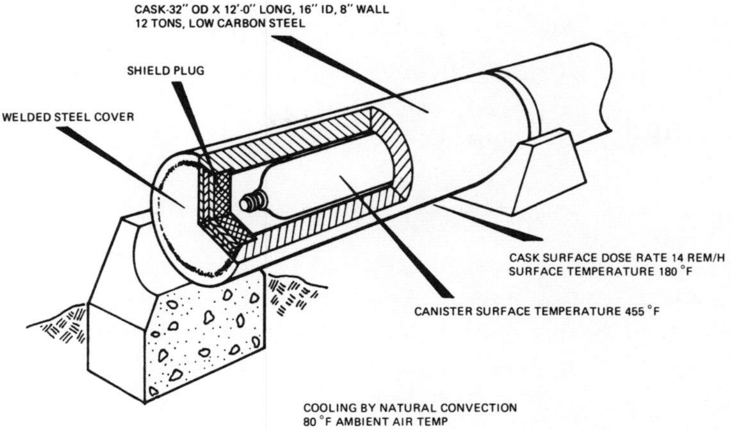

FIGURE C-1
Diagram of the preliminary AEC sealed storage-cask concept—unshielded storage unit on concrete saddles.

peratures within the storage array are expected to be about 180 °F. Under these conditions, the surface temperature of the waste canister will be 455 °F.

OPERATING REQUIREMENTS

Land requirements will be about 2,300 acres by 2010. Personnel requirements will be about 86 in 1983, 334 in 2010, and 76 in 2011. Water usage will be about 74 million gallons per year in 2010 and electrical usage 3,000 kilowatts.

TECHNOLOGY

The principles of heat transfer by natural convection from metal surfaces to ambient air, as applied in this concept, are well known and widely used.

The fabrication of heavy steel casks for at least a 100-year life has not been demonstrated. Engineering verification work will be required in the application of existing welding and nondestructive testing technology to remote operations. High quality-assurance standards and nondestructive tests must be developed and demonstrated to assure the long-term, leak-free integrity of a cask.

MATERIALS USAGE

Approximately 980,000 tons of steel, of which 920,000 tons will be used for fabrication of casks, will be used through 2010. Concrete usage through 2010 will be about 90,000 cubic yards.

SAFETY CONSIDERATIONS

This concept utilizes passive cooling and a high-integrity cask that is considered capable of withstanding postulated accidents. The concept assumes total reliance on the cask to confine the contents of a waste canister. Reliable cask-fabrication and testing techniques, or means to verify continued canister and/or cask integrity, will require further evaluations. The high radiation field in storage areas requires shielded vehicles and strict administrative procedures for personnel protection.

ESTIMATED COSTS (1973 DOLLARS)

Total estimated program costs through 2010 are $1.3 billion, including $1.0 billion for casks. Operation through 2110 is estimated to cost an additional $0.3 billion.

APPENDIX D
SUMMARY DESCRIPTION OF WATER BASIN CONCEPT—NO OVERPACK

STORAGE

In the water basin concept (WBC) for storage of high-level radioactive waste, the bare canisters (Figure D-1) are stored in stainless steel-lined concrete basins. Each basin has a nominal capacity of 500 canisters stored in double-tiered racks for support. The basins are filled with demineralized water, which cools the waste and provides transparent radiation shielding and a confinement barrier.

Heat is removed by the constant circulation of the basin water in direct contact with the canisters. It is transferred to a secondary cooling loop and rejected to the atmosphere from standard, open, forced-draft cooling towers located just outside the storage building. A diagram of the system is shown in Figure D-2.

Each basin is equipped with a pump and heat-exchanger system to maintain basin water temperature below 120 °F, and the canister surface temperature below 128 °F. Water purity is maintained at less than 10 ppm chloride by passing a portion of the circulated cooling water through a filtration and demineralizer system. The radiation dose rate at the surface of the water in the basins is estimated to be less than 1 mrem/h.

Prepared by the U.S. Atomic Energy Commission.

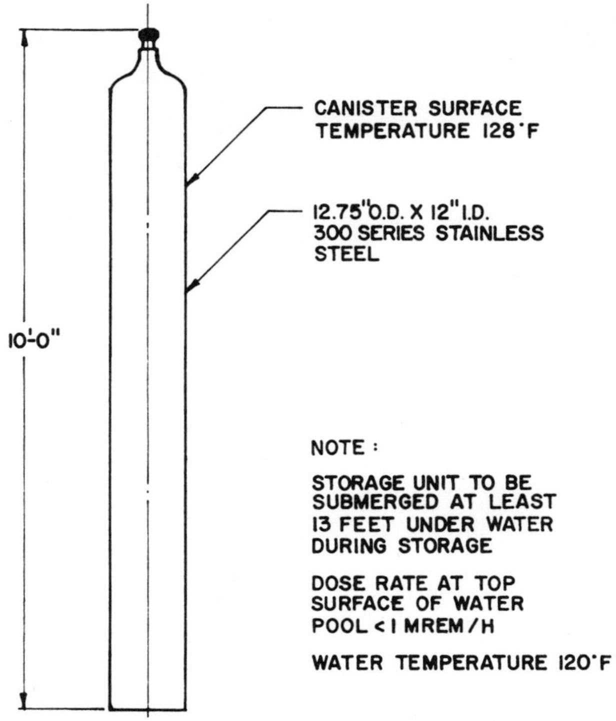

FIGURE D-1
Diagram of the preliminary AEC water basin concept storage unit.

OPERATING REQUIREMENTS

Land requirements will be about 100 acres by 2010. Personnel requirements will be 89 in 1983, 251 in 2010, and 155 in 2011. Water usage will be about 1.5 billion gallons in 2010 and electrical usage 10,400 kilowatts.

TECHNOLOGY

The handling and storage of radioactive materials under water is a standard method of operation in the nuclear industry. For over 25 years, reactor fuel elements have been routinely discharged to and stored in water basins. Water basins are also utilized for handling radioactive sources at research facilities. The storage of high-level waste canisters in water basins for 100 years is an extension of present technology. Care in the design and operation of systems for chemical adjustment and control of basin water quality will be required to minimize corrosion effects during long-term storage.

MATERIALS USAGE

Approximately 18,000 tons of carbon steel and 5,600 tons of stainless steel will be used through 2010. Concrete usage through 2010 will be about 100,000 cubic yards.

SAFETY CONSIDERATIONS

The WBC utilizes a cooling system that requires continuous surveillance and occasional equipment maintenance. Equipment for close control over water chemistry is needed to minimize corrosion of canisters and basin liners. The basins and liners provide physical protection of canisters except for impact from above. Basin water serves as an additional confinement barrier and heat sink.

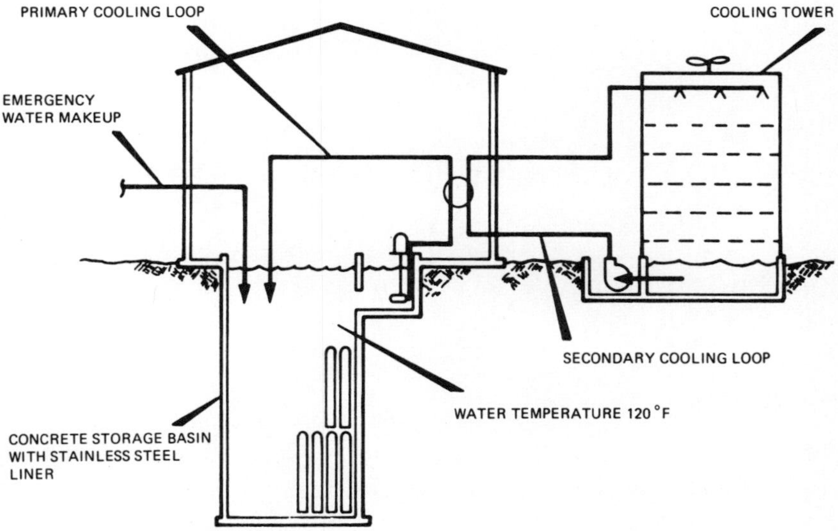

FIGURE D-2
Diagram of the preliminary AEC water basin concept facility.

Canister failure in storage is an expected, infrequent occurrence. Design features are required to allow recovery from such failures and decontamination of basin water.

ESTIMATED COSTS (1973 DOLLARS)

Total estimated program costs through 2010 are $0.45 billion (no overpacks or casks required). Operation through 2110 would cost an additional $1.0 billion.

APPENDIX E
SUMMARY DESCRIPTION OF LIQUID STORAGE OF HIGH-LEVEL WASTE

STORAGE

In the liquid storage concept for high-level radioactive waste, the form of the waste is an acidic liquid nitrate. The waste is stored underground in a multilayer containment vessel (see Figures E-1 and E-2). Primary containment is provided by a 304L stainless steel tank with a wall thickness that varies from 1/2 inch at the bottom to 7/16 inch at the top. The tank is approximately 50 feet in diameter by 20 feet high, with a rated capacity of 300,000 gallons, and a design life of 25–30 years. Secondary containment is provided by a 304L stainless steel liner, which encloses the sides and bottom of the tank. The 1/4-inch-thick liner is approximately 60 feet in diameter by 18 feet high. Enclosing the tank and liner is a concrete vault with a 60-foot internal diameter and 26-foot internal height. The floor of the vault is 4 feet thick, the side walls are 3 feet thick, and the concrete dome over the top is 5 1/2 feet thick.

Tank cooling is provided by cooling coils of 2-inch 304L stainless steel pipe and by cooling towers. The cooling system is rated at 37×10^6 Btu/h for each tank.

Prepared by the U.S. Atomic Energy Commission.

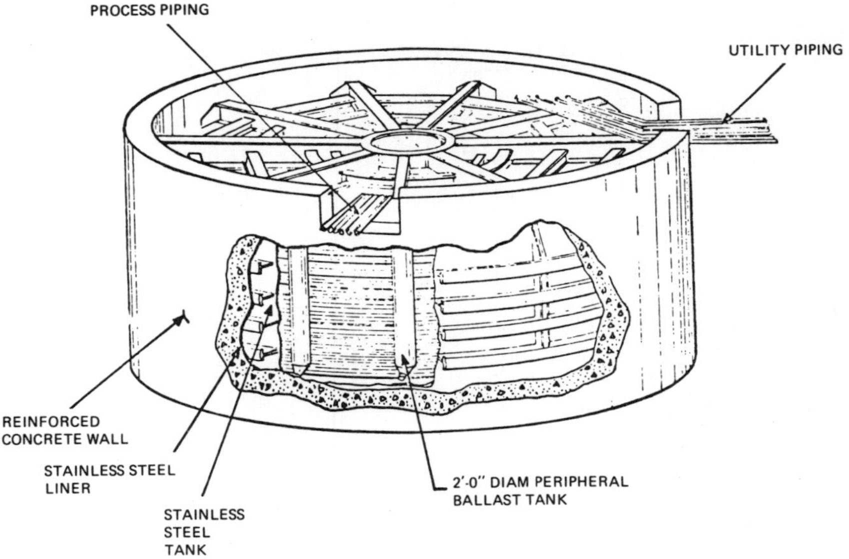

FIGURE E-1
Diagram of the preliminary AEC high-level liquid waste storage tank.

OPERATING REQUIREMENTS

Operation of the liquid waste storage facility would require 65 tanks by the year 2010. Land requirements for these tanks and associated cooling equipment would be 100 acres. It is anticipated that 2 billion gallons of water and 14,000 kilowatts of electrical power would be consumed per year of operation by 2010 and a staff of about 150 people would be required.

TECHNOLOGY

Storage of high-level wastes as liquid has been used for over 25 years at Hanford and is also in use at Savannah River. To protect against leakage, the more advanced tanks are double contained, with the tanks located inside concrete vaults.

In the United Kingdom liquid waste storage in stainless steel, double-contained tanks during the past 20 years has yielded much information on physical and chemical properties of materials, as well as corrosion data. In the United States, both the Nuclear Fuel Services' plant at West

Valley, New York, and the Allied Gulf Nuclear Fuel Plant at Barnwell, South Carolina, employ tank storage. The Barnwell tanks are based on the United Kingdom's design for nonboiling liquid wastes and contain advanced engineering features.

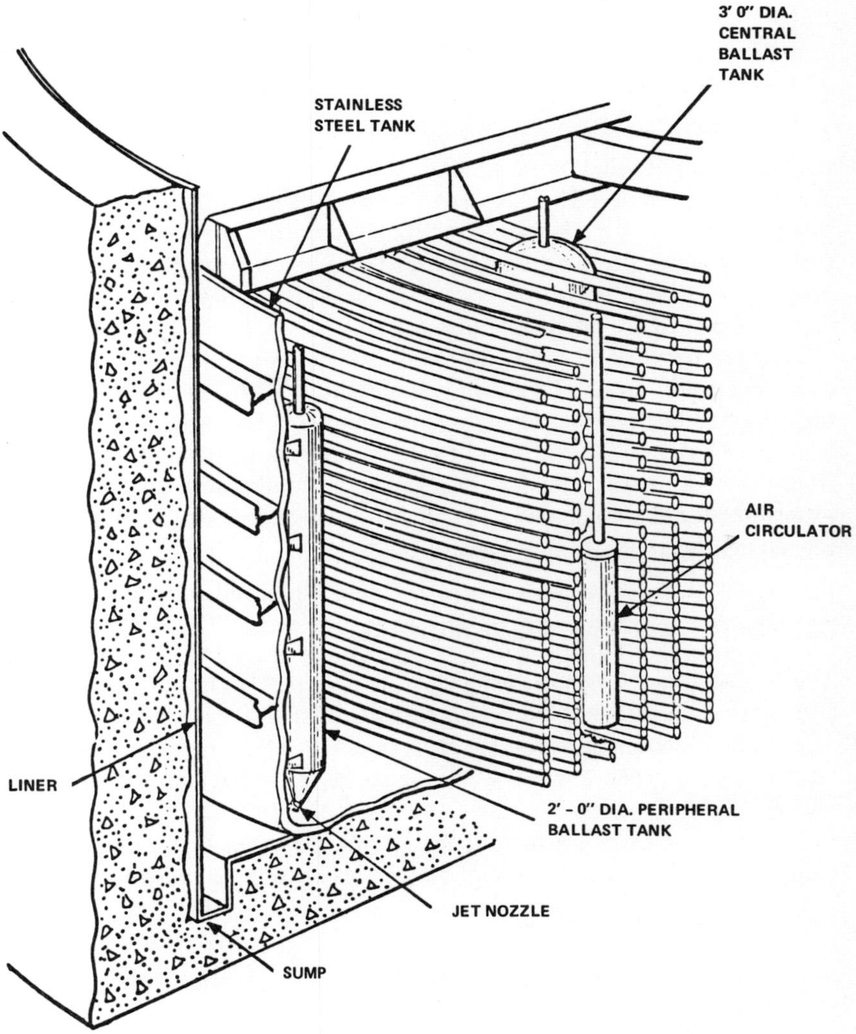

FIGURE E-2
Detail of the preliminary AEC high-level liquid waste storage tank.

MATERIALS USAGE

About sixty-five 300,000 gallon tanks, as previously described, would be required to store all the liquid waste generated by commercial fuel processors by the year 2000. Construction of these tanks would require about 10,000 tons of stainless steel and 120,000 cubic yards of concrete. Since the design life of each tank is 25–30 years, at least 220 tanks would have to be built during the course of 100-year storage.

SAFETY CONSIDERATIONS

To preclude loss of confinement of the radioactive liquid wastes in the tanks, the double-shell concept tank within a steel-lined vault provides an additional barrier. This concept also permits monitoring for leaks of the primary shell. Dry air may be circulated within the vault between the tank and liner to provide a measure of corrosion control and thus prolong the active life of the tanks. The tank design itself allows 1/8 inch of material for corrosion allowance.

Low temperature (120 °F) operation of the nonboiling liquid minimizes corrosion and metal stresses. Maintaining an internal cooling-coil pressure higher than the tank pressure eliminates the potential for leakage of the radioactive liquid to the cooling water system. In addition, the cooling water system design contains individual supply and return header units, each of which can be completely isolated from the system by valves in the event of leakage.

ESTIMATED COSTS (1973 DOLLARS)

The total estimated cost to provide a facility to store liquid wastes generated through the year 2000 is estimated to be $200 million. Because of the 25–30 year design life for storage tanks, a new facility must be built every 30 years, and the wastes must be transferred to the new tanks.

The estimate does not include land costs; however, this must be included if the facility is not located on an existing AEC controlled site.

APPENDIX F
RETRIEVABLE SURFACE-STORAGE FACILITY AND SEALED STORAGE-CASK CONCEPT—SUPPLEMENTARY INFORMATION

Included are temperature data for canisters of 10-year-old waste from pressurized water reactor-uranium (PWR-U) type fuel exposed to 33,000 MWd/t sealed in a 2-inch-thick steel cask, with a 6-inch-wide, air-cooling annulus between the cask and a 38-inch-thick concrete shield.

Table F-1 lists the normal operating temperatures for 5- and 10-kilowatt canisters with natural draft, air-flow cooling, and the abnormal condition temperatures estimated with no air flow. Limiting temperatures are listed for the various storage unit components, and are defined in Table F-2. Table F-3 shows the estimated time to reach limiting temperatures under buried conditions for 5- and 20-kilowatt canisters in a sealed storage cask.

The maximum heat content of a 12-inch diameter × 10-foot long canister is about 10 kilowatts, to maintain acceptable temperatures under the no-air-flow condition. A canister in a sealed cask storage containing less than 5 kilowatts could be buried indefinitely without exceeding the limiting temperatures. Further optimization studies and heat transfer demonstrations will be required. Methods and assumptions used for temperature calculations of a sealed storage cask are also included.

Prepared by the U.S. Atomic Energy Commission.

TABLE F-1

Sealed Storage-Cask Component Temperature Data

		With Air Flow Temperature, °F			No-Air-Flow[c] Temperature, °F		
		Operating	Limiting	Factor[d]	Operating	Limiting	Factor[d]
5 kW[a]	Canister	470	800	1.7	810	1,500	1.8
	Cask	245	550	2.2	700	1,200	1.7
	Concrete, inside surface	155	500	3.2	650	900	1.4
	Concrete, outside surface	84	500	5.9	120	900	7.5
10 kW[b]	Canister	650	800	1.2	1,300	1,500	1.2
	Cask	390	550	1.4	1,200	1,200	1.0
	Concrete, inside surface	255	500	2.0	1,190	900	0.8
	Concrete, outside surface	95	500	5.2	135	900	6.7

NOTE: All data is for PWR-U-type waste, 2-inch-thick cask, 6-inch annulus.
[a]For 5 kW, Air flow = 3 fps, 600 CFM, Ambient Air = 80 °F, Exit Air = 115 °F.
[b]For 10 kW, Air flow = 4.2 fps, 830 CFM, Ambient Air = 80 °F, Exit Air = 190 °F.
[c]For the no-air-flow data, the limiting temperatures used for the canister and the cask are based on 240 hours.
[d]Factor = Limiting temperature/operating temperature.

TABLE F-2
Sealed Storage-Cask Concept Material Temperature Limitations in Air

	Normal Operation[a]	Abnormal Operation[b]
Concrete: Structure	500 °F[c]	900 °F[d]
Shield		
Carbon steel: Cask	600 °F[e]	1200 °F[f]
Stainless steel: Canister	800 °F[g]	1500 °F[h]
Overpack		

Bases: [a]Maximum allowable temperatures during 100 year design life that will not compromise containment integrity.
[b]Maximum allowable temperature for up to one year where immediate containment integrity would not be impaired, but integrity could not be assured for 100 years.
[c]Twenty percent loss of strength.
[d]Extensive cracking with significant loss of strength; concrete would require replacement.
[e]Maximum oxidation rate of 2 mils/yr.
[f]Upper limit of carbon steel usefulness in air; however, under no-air-flow conditions, oxidation rate would be low. There would be some loss of ductility due to carbide spheroidation.
[g]Grain boundary attack by cesium oxide in calcine waste would be accelerated above 850 °F.
[h]Mechanical and fatigue properties decreased significantly.

METHOD AND ASSUMPTIONS FOR TEMPERATURE CALCULATIONS FOR A SEALED STORAGE CASK

Method

Temperature-profile analysis was by the finite difference method using a standard digital computer code for diffusive processes. The physical system was modeled using a lumped-geometry network representative of the conductive, convective, and radiative heat flow paths and the admittances found in a storage unit. The method yields both transient and steady-state temperatures.

Estimates of the temperatures of interior surfaces (canister and interior-cask surface) considered heat transfer by radiation and natural convection only. Radiant-interchange factors were taken as a function of surface geometries and emissivities, and the convective heat-transfer coefficient used was that for an enclosed space between two vertical flat plates. End effects were similarly treated although a no-flux bottom was assumed. Further, it was assumed that there is no significant gas film inside the canister.

Between cask and shield, temperatures were estimated by radiative and convective energy balances on the surfaces in the enclosure and the

TABLE F-3

Estimated Time to Reach Limiting Temperatures for Overpacked Canister Buried in Ground (2 examples)

Situation 1:	
Heat generation	5 kW
Canister diameter	12 in.
Canister length	10 ft
Burial position	Vertical
Burial depth	Top–5 feet below grade Midplane–about 10 feet below grade
Overpack	
Diameter	19-inch O.D. (2-inch-thick wall)
Length	11 ft
Material	Carbon steel Maximum temperature carbon steel can attain and still have some structural strength for some time: 1400 °F Carbon steel melting temperature about 2700 °F
Time for overpack to reach 1400 °F–about 8 weeks	
Maximum overpack temperature reached is less than 1700 °F after about 1.5 years	
Situation 2:	
Heat generation	20 kW
Canister diameter	24 in.
Canister length	10 ft
Overpack diameter	30 in.
Overpack length	11 ft
Time for overpack to reach 1400 °F–about 2 weeks	
Time for overpack to reach 2700 °F–about 1 month	

convective air stream. A temperature profile for air in the annulus was established with inlet temperature assumed to be equivalent to the mean August Hanford day. Natural convection heat-transfer coefficients were used for annulus surfaces, a conservative assumption in the normal operation case. The annulus air-flow rate was determined by equating the buoyancy of the air in the gap due to heating to the hydraulic resistance of the flow path. A total loss of air flow was assumed in the air-flow blockage case.

The assembly was assumed to be in vertical orientation. Solar heating of the exterior surface was not considered. Diurnal variation of the ambient-air temperature was not treated because of the conservatively high average value chosen.

Assumptions

The major assumptions used for these calculations are indicated below.

1. Assembly bottom is no-flux surface.
2. Inlet air temperature (80 °F) is that of mean August Hanford day.
3. Effects of solar heating are not significant.

The materials properties used in these calculations are as follows:

Canister
Material of construction: type 304L stainless steel
Thermal conductivity: 10.0 Btu/h-ft-°F
Emissivity: 0.45
Heat capacity: 0.12 Btu/lb-°F
Density: 500 lb/ft^3

Cask
Material of construction: carbon steel
Thermal conductivity: 26.0 Btu/h-ft-°F
Emissivity: 0.6
Heat capacity: 0.12 Btu/lb-°F
Density: 490 lb/ft^3

Shield
Material of construction: concrete
Thermal conductivity: 0.5 Btu/h-ft-°F
Emissivity: 0.7
Heat capacity: 0.2 Btu/lb-°F
Density: 143 lb/ft^3

Waste
Heat generation: 5 kW total (0.66 kW/ft^3)
Thermal conductivity: 0.2 Btu/h-ft-°F
Heat capacity: 0.2 Btu/lb-°F
Density: 81 lb/ft^3

Air (per inlet)
Temperature: 80 °F
Thermal conductivity: 0.015 Btu/h-ft-°F
Heat capacity: 0.24 Btu/lb-°F
Density: 0.072 lb/ft^3

Burial Earth
Thermal conductivity: 0.3 Btu/h-ft-°F
Soil thermal diffusivity: 0.0083 ft^2-°F
Heat capacity: 0.2 Btu/lb-°F
Density: 180 lb/ft^3

APPENDIX G

June 3, 1974

Dr. Dixy Lee Ray
Chairman
U.S. Atomic Energy Commission
Washington, D.C. 20545

Dear Chairman Ray:

In a letter of November 7, 1972, the AEC requested the Committee on Radioactive Waste Management (CRWM) for advice "concerning the AEC's concepts for the safe storage of high-level solidified waste for interim storage in an engineered storage facility at or near the surface of the earth."

The response of the CRWM was to establish the Panel on Engineered Storage, which over the past 15 months has evaluated the pertinent information on management of high-level radioactive wastes that will be generated by nuclear power reactors. A comprehensive study of the management of these wastes has been undertaken, and a report on this study is in preparation. As a consequence of this study, the Panel has concluded that retrievable surface storage of high-level radioactive wastes is an essential element in the development of the overall radioactive waste management system and one which has the potential to (a) provide for public safety by multiple barriers against dispersal of radionuclides and against exposure of persons to radioactivity, (b) have minimal environmental effects, (c) be implemented as soon as needed, and (d) have minimal impact on nuclear power costs.

Thus, the Panel, with the concurrence of the CRWM, recommends that—

1. detailed engineering plans be developed for a Retrievable Surface Storage Facility (RSSF). A RSSF for an interim period (up to 100 years) is understood to be a provisional measure to be used while procedures for long-term storage or ultimate disposal of radioactive wastes (e.g., in deep geologic strata) are being developed, proven, and accepted;
2. a Sealed Storage Cask (SSC) facility be built and put into operation on commercially produced high-level radioactive wastes. An optimized version of the sealed cask concept would be simple and economically feasible, and would also afford considerable flexibility in development, implementation, and incorporation of future improvements;
3. the U.S. Atomic Energy Commission continue developing concepts of fractionation, solidification, containerization, and retrievability of high-level radioactive wastes, with an emphasis on a systems approach, leading to ultimate disposal.

Since a RSSF is designed to serve for an interim period, it must, by its very nature, be replaced in the future by an alternative method of managing radioactive wastes. We believe that the search for safe methods of storing these wastes for the long term or for their ultimate disposal must continue.

We understand that budgetary considerations require the Commission to act on the subject of a RSSF at an early date. It is for this reason that we are providing these recommendations at this time.

Sincerely,

JOHN C. FRYE
Chairman, Committee on
Radioactive Waste Management

APPENDIX H

February 12, 1975

Dr. Robert C. Seamans, Jr.
Administrator
U.S. Energy Research and Development Administration
Washington, D.C.

Dear Dr. Seamans:

The Committee on Radioactive Waste Management (CRWM) of the NAS-NAE-NRC appointed a Panel on Transportation* to review the transportation of radioactive wastes. The purpose of the Panel, as an advisory group to the CRWM was (1) to determine the probabilities and consequences of transportation accidents, (2) to estimate the potential risks of transporting radioactive wastes (including handling and packaging), and (3) to attempt to evaluate these risks either on an absolute basis or by comparing them to other risks involved in the management of radioactive wastes. (The risks here may be defined in terms of the potential for injury and damage to human beings, the environment, and property from the release of, or exposure to radioactive waste materials.)

The Panel considered only the following topics:

1. The transportation of high-level solid radioactive wastes, particularly those from reprocessing plants, excluding shipments of spent fuel elements or any form of liquid wastes.
2. The adequacy of present or proposed designs for containers and

*Clarke Williams, *Chairman*; Thurber C. George; Alfred W. Grella; John W. Langhaar; Herbert M. Parker; G. Hoyt Whipple; and Cyrus Klingsberg, *Executive Secretary*.

packaging in terms of radiation protection and the prevention of unacceptable exposure to individuals, population groups, or other forms of life in the environment.

3. The adequacy of the regulations and standards to ensure that items 1 and 2 will be satisfactory.

4. The adequacy of the requirements of the quality assurance program and administrative procedures for quality control for the design, fabrication of containers, and preparation of packages for shipment.

5. The adequacy of the operational provisions cited above over the next few decades, both in terms of the projected increase in volume of shipments of radioactive wastes and in terms of public exposure or any other changes or improvements that may be required over this period.

In December 1972, the AEC published *Environmental Survey of the Transportation of Radioactive Materials to and from Nuclear Power Plants* (WASH-1238). The Panel believes that the methodology developed in the Environmental Survey is capable of being modified to provide a fair estimate of the risks involved in transporting high-level solid radioactive waste. However, until a thorough analysis of the whole nuclear fuel system from mine to ultimate disposal is completed, a final judgment of the relative importance of transportation cannot be made.

Conclusions are reached in the AEC Environmental Survey as to the risk from transporting high-level radioactive fuel elements on the exposed segment of the public, as well as transportation workers and employees in the industry. This exposure is at least several orders of magnitude lower than the most restrictive acceptable level, as contained in the Radiation Protection Guides recommended by the Federal Radiation Council (criteria are currently available from the Environmental Protection Agency). There is some reason for thinking that the high-level solid radioactive wastes would offer even less hazard in shipment than the spent fuel elements because of their physical and chemical characteristics when processed for shipment to the storage site. However, by using the methodology of the Environmental Survey, even if the analysis of the potential risk from the transportation of the radioactive waste were to be higher by a factor of ten, the total adverse effect would still be well within specified guidelines.

The AEC Environmental Survey developed a methodology for estimating the impact on employees, transportation workers, and the general public from radiological effects, as well as conventional accidents under normal conditions of transportation by truck, rail, and barge of nuclear fuel to and from power plants. An analysis was made in the Environmental Survey of the frequency and severity of accidents occurring during transportation, the likelihood of release of radioactivity, and the estimated resultant exposures due to the accidents. The Academy Panel found that, on the whole, the methodology, although developed for transportation of nuclear fuel, was responsive to their concern expressed earlier for a systematic attack on the waste transportation problem, as indeed spent fuel elements are a highly radioactive solid material. Our finding was that the Environmental Survey methodology could be extended to the transportation of high-level solid radioactive waste from processing plants to interim and final storage areas.

Since the Panel finds the exposures developed in the Environmental Survey acceptable for shipments from a single reactor, it would appear that radiologically equivalent shipments of high-level solid radioactive waste would also be acceptable.

The Panel believes that, although the probability of more severe accidents than those considered in the Environmental Survey is extremely small, experience has shown that extremely unlikely events do occur. Since no one can be totally convinced that a very serious accident will never occur, one such incident should be factored into the total radiation exposure in terms of man rems over the next 40-year period. Waste management practices should be planned so that, if one such major radiological accident ever occurs, a complete moratorium on transportation of high-level radioactive material could be maintained for a sufficiently long period to permit orderly study and correction of the defect. The requirement of a moratorium might well affect the design criteria for casks that ultimately involve cost–benefit balancing and must be related to the whole system. As an example, if more shielding or structural impact-resistant material is required per cask so that less radioactive waste is contained in each cask, the increased number of shipments may well result in greater environmental costs, including increased nonradiological hazards. In fact, from the limited perspective available to the Panel, it seems likely that the overwhelming risk in transportation is that of direct death and injury and damage to property from conventional traffic accidents totally unrelated to radiation.

In addition, the Panel felt that, in order to ensure the designed integrity of shipping casks, the greatest care must be taken in their manufacture, testing, use, and maintenance. Present experience with nuclear power plant components has shown that, in many cases, the quality-assurance programs have not been adequately carried out. The Panel believes that ERDA must continue to strive for the highest quality-assurance programs. The Panel concludes that the transportation of radioactive materials is not a major factor in the total hazards associated with the nuclear power system. However, this conclusion is supportable only if the highest standards of care are applied to all aspects of waste preparation and transportation.

In August 1974, our Panel wrote a letter-report to the CRWM. This internal working document is the final effort of the Panel and is not a formal Academy report. Nevertheless, we believe that some of the findings of the Panel on Transportation will be useful to your transportation program. Consequently, the Committee on Radioactive Waste Management has reviewed this letter-report and concurs in the following recommendations of its Panel on Transportation:

1. that ERDA develop information on type, quantity, form, and physical characteristics of high-level solid radioactive wastes expected over the foreseeable future (20–40 years), that research be continued to develop transportable wastes that will be affected least by exposure to the environment (lowest solubility, highest fire resistance, maximum mechanical integrity to prevent dissemination of particulates, etc.), that ERDA extend the methodology of the Environmental Survey to specifically determine the effects of the transport of these wastes on the employees, transport workers, and public, with the ultimate objective of determining whether possible changes in NRC and DOT regulations on such shipments are needed. This study should include the consideration of the effect of a more severe accidental release than is considered in the present Environmental Survey, and the effect of the increasing annual rate of future shipments to a storage site.

2. that the program for testing casks for shipping radioactive materials and improving the design where possible be continued and extended to cover casks specifically designed for shipment of high-level solid radioactive waste.

3. that ERDA require and assure that a high level of quality be made effective in the manufacture, testing, use, and maintenance of shipping containers and vehicles.

4. that the total systems analysis of the whole nuclear power system be made so that recommendations for the future management of radioactive wastes can be formulated in relation to other parts of the system.

Sincerely yours,

JOHN C. FRYE
Chairman, Committee on
Radioactive Waste Management

GLOSSARY AND ABBREVIATIONS

actinide elements Elements in the periodic table of the elements with atomic numbers 90 to 103. All are radioactive.

ACVC Air-cooled vault concept

calcine A powdered or friable product (e.g., metal oxide) of roasting

CRWM Committee on Radioactive Waste Management

disposal Isolating the radioactive waste permanently in a form and manner that precludes retrieval

fission products Any of the primary fragments that result from the fission of heavy nuclei, together with their radioactive decay products. About 200 different stable and radioactive nuclides have been identified as fission products.

fps Feet per second

gigawatt Unit of energy equal to 10^9 watts

high-level radioactive waste (a) Aqueous waste resulting from the first cycle of a reprocessing operation, or (b) products from the solidification of high-level liquid wastes, or (c) irradiated fuel elements, if discarded without processing.

interim storage Not exceeding 100 years (as distinct from long-term storage)

long-term storage Storage for an indefinite length of time, in a retrievable fashion (see disposal)

μm Micrometer, a unit of length equal to 10^{-6} meters

MCi Megacuries = 10^6 curies of radioactivity

MWde 10^6 Watts (electrical) per day

MWdth Units of work equal to 10^6 watts (thermal) per day

mrem/h Millirem/hour, unit of dose-equivalent rate

n/s cm^2 Unit of neutron flux in neutrons per second per square centimeter

overpack An outer metal container (i.e., a second canister) that is sealed to provide double confinement of the radioactive waste

psi Pounds per square inch

PWR-U-type waste Pressurized water reactor-uranium type waste

rem A unit of absorbed dose of ionizing radiation in biological matter

RSSF Retrievable surface-storage facility

SSCC Sealed storage-cask concept

t Metric ton, equal to 1000 kilograms or 2200 pounds

transuranics An acceptable condensation of the terms *transuranium elements* or *transuranic elements.* Elements heavier than uranium in the periodic table, i.e., elements with atomic numbers above 92. All transuranic elements are radioactive and do not occur in nature except for very small amounts of plutonium, which result from the interaction of cosmic ray neutrons with uranium.

WBC Water basin concept

WBC-OP Water basin concept with overpack